NUREG-1478
Rev. 2

Operator Licensing Examiner Standards for Research and Test Reactors

Manuscript Completed: April 2007
Date Published: June 2007

Division of Policy and Rulemaking
Office of Nuclear Reactor Regulation
U.S. Nuclear Regulatory Commission
Washington, DC 20555-0001

SUMMARY OF SIGNIFICANT CHANGES

Note:	All standards have been revised to remove the name of the branch and division responsible for the administration of operator licensing for research and test reactors. This was done to improve efficiency by reducing the administrative burden on the staff to revise this standard when the name of a branch and/or division occurs. In addition the standards have been modified to a two-column format.
ES-101N	Added reference to new 700-series senior reactor operator limited to fuel handling (LSRO).
ES-102N	Removed reference to Regulatory Guide 1.114, "Guidance to Operators at the Controls and to Senior Operators in the Control Room of a Nuclear Power Unit," issued May 1989, which is applicable to power reactors only.
ES-201N	Changed time for issuance of confirmatory letter to facility from 120 days before scheduled examination date to 90 days. Deleted reference to the use of contract examiners. Deleted reference to submitting preliminary license applications 30 days before examination start date.
ES-202N	Deleted reference to submitting preliminary license applications 30 days before examination start date. Added further clarification concerning the review of the NRC form 396 to check for the actual guidance used during the medical examination.
ES-203N	No significant changes.
ES-204N	Moved procedures for reviewing license renewal applications from ES-601N and ES-602N to this new standard.
ES-301N	Moved portions of this standard that dealt with operating test administration to ES-302N and portions that dealt with operating test grading to ES-303N. Renamed Category B from Control Room Systems (B.1) and Facility Walkthrough (B.2) to Facility Walkthrough (B). Added definitions for Reactivity evolutions and Major Transients (with entry into emergency plan) to Category C. Modified reactor operator (RO) and senior reactor operator (SRO) competency summaries to better reflect those necessary and able to be tested for operators at research and test reactors. Revised table containing systems for operating tests to better reflect systems at research and test reactors. Deleted Transient/Event Checklist. Moved Competency Review Worksheets from ES-301N to ES-303N. Modified scope of examination administered based on facility type (a new table defines facility types).
ES-302N	Moved portions of ES-301N that dealt with operating test administration into this standard and deleted redundancies. Modified examination administration requirements to reflect different facility types.

ES-303N	Moved Competency Review Worksheets from ES-301N to ES-303N. Revised the Competency Review Worksheets to better reflect items important to an RO and an SRO at a research and test reactor and to better reflect the methods of examination possible at a research and test reactor.
ES-401N	Changed written examination requirements based on facility types introduced in ES-301.
ES-402N	Deleted Attachment 3, "Policies for Facility Review of Written Examinations," which the body of standards covered.
ES-403N	Deleted reference to examinations prepared and graded by contract examiners. Moved quality assurance reviews to ES-501 which already covered issue under examination reviews.
ES-501N	Revised Form ES-501N-2 to reflect written examination containing three sections. Deleted Form ES-501N-3 which dealt with contract examiner evaluations.
ES-502N	No major changes.
ES-601N	Standard rewritten to flow from preexamination activities to examination administration activities to postexamination activities.
ES-602N	Moved sections on maintaining medical standards for licensees and procedure for denying an application for license renewal to ES-204N.
ES-701N	New standard added to cover general items related to administration of LSRO examinations at permanently shutdown research reactors.
ES-702N	New standard added to cover preexamination activities related to administration of LSRO examinations at permanently shutdown research reactors.
ES-703N	New standard added to cover examination administration activities related to administration of LSRO.
ES-704N	New standard added to cover postexamination activities related to administration of LSRO.

ABBREVIATIONS

ADAMS	Agencywide Documents Access and Management System
ANSI/ANS	American National Standards Institute/American Nuclear Society
BC	branch chief
BC/PD	branch chief or program director with program responsibility
CE	chief examiner assigned to an examination
CFR	*Code of Federal Regulations*
EPIP	emergency plan implementing procedures
ES	examiner standards
FSAR	final safety analysis report
GED	General Education Development test
K/A	knowledge and ability
KSAs	knowledge, skills, and abilities
LSRO	senior reactor operator limited to fuel handling
NRC	U.S. Nuclear Regulatory Commission
OLA	licensing assistant responsible for research and test reactors operator licensing docket files
PD	program director
QA	quality assurance
RO	reactor operator
RTROL	research and test reactor operator licensing
SRO	senior reactor operator
SRO-I	senior reactor operator-instant
SRO-U	senior reactor operator-upgrade

CONTENTS

EXAMINER STANDARD 101N
PURPOSE AND FORMAT OF NUREG-1478

A. Purpose

The research and test reactor operator licensing (RTROL) examiner standards establish the procedures and practices for administering initial and requalification written examinations and operating tests to license applicants and operators pursuant to Title 10, Part 55, "Operators Licenses," of the *Code of Federal Regulations* (10 CFR Part 55).

These standards describe the provisions of the Atomic Energy Act of 1954, as amended and the regulations of 10 CFR Part 50, "Domestic Licensing of Production and Utilization Facilities," and 10 CFR Part 55 on which the program is based. These standards also provide for equitable and consistent administration of examinations to all applicants and licensed operators at research and test reactor facilities.

B. Format

Each standard describes procedures and practices for a particular aspect of the Research and Test Reactor Operator Licensing Program. Each standard is assigned a three-digit number, and related standards are grouped together for ease of reference. All the standards beginning with the same digit apply to related aspects of the program as follows:

(1)   general
(2)   preexamination activities
(3)   operating tests
(4)   written examinations
(5)   postexamination activities
(6)   requalification examinations
(7)   senior reactor operator limited to fuel handling (LSRO) licenses

EXAMINER STANDARD 102N
APPLICABLE REGULATIONS AND PUBLICATIONS

A. Purpose

This standard lists the U.S. statute and the regulations of the U.S. Nuclear Regulatory Commission (NRC) that establish the requirements for the conduct of RTROL examinations and the procedures for their administration. It also identifies the regulatory guides, NUREG reports, and other published guidance intended to implement the regulations and the American National Standards Institute/American Nuclear Society (ANSI/ANS) standards that may provide additional guidance.

B. Statutes

1. Atomic Energy Act of 1954

Section 107 of the act, as amended (42 U.S.C. 2137), requires the NRC to prescribe uniform conditions for licensing individuals as operators of production and utilization facilities, to determine the qualifications of these individuals, and to issue licenses to such individuals.

C. Regulations

1. 10 CFR Part 55

The implementing regulation for licensing reactor operators (ROs) and senior reactor operators (SROs) is found in 10 CFR Part 55. This regulation establishes the requirements and the regulatory basis for licensing and requalifying operators.

2. 10 CFR Part 50

As required by 10 CFR 50.34(b)(8), the final safety analysis report (FSAR) must include a description of the operator requalification program. This description forms the basis for the inspection, audit, and approval of requalification programs.

10 CFR 50.54(i-1), states "Within three months after issuance of an operating license, the licensee shall have in effect an operator requalification program, which as a minimum, must meet the requirements of § 55.59(c). Notwithstanding the provisions of § 50.59, the licensee may not decrease the scope of its approved requalification program without authorization from the Commission."

The regulations found in 10 CFR 50.54(k)–(m) restrict control manipulations to licensed operators and conditions of all facility licenses issued under 10 CFR Part 50.

As required by 10 CFR 50.74, "Notification of Change in Operator or Senior Operator Status," facility licensees must notify the Commission within 30 days if the status of a licensed RO or SRO changes.

3. 10 CFR Part 2

The regulations in 10 CFR Part 2, "Rules of Practice for Domestic Licensing Proceedings and Issuance of Orders," govern the conduct of all proceedings under the Atomic Energy Act of 1954, as amended, and the Energy Reorganization Act of 1974 for (a) granting, suspending, revoking, amending, or taking other action with respect to any license, (b) imposing civil penalties, and (c) public rulemaking.

The applicant's right to demand a review of a proposed license denial, including appeal and hearing rights, is established in 10 CFR 2.103(b)(2).

Subpart C, "Rules of General Applicability: Hearing Requests, Petitions to Intervene, Availability of Documents, Selection of Specific Hearing Procedures, Presiding Officer Powers, and General Hearing Management for NRC Adjudicatory Hearings," of 10 CFR Part 2 governs all adjudications initiated by the issuance of an order to show cause, an order designating the time and place of a hearing requested by a person charged with a violation, and a notice of hearing.

Subpart L, "Informal Hearing Procedures for NRC Adjudications," governs proceedings for the granting, renewal, or licensee-initiated amendment of an RO or SRO.

4. 10 CFR Part 9

The regulations in 10 CFR Part 9, "Public Records," prescribe the rules governing the NRC's public records that relate to any proceeding subject to 10 CFR Part 2.

Subparts A, "Freedom of Information Act Regulations," and B, "Privacy Act Regulations," of 10 CFR Part 9 describe and implement the requirements for balancing the public's right to information under the Freedom of Information Act and the NRC's responsibility to protect personal information under the Privacy Act.

Subparts C, "Government in the Sunshine Act Regulations," and D, "Production or Disclosure in Response to Subpoenas or Demands of Courts or Other Authorities," of 10 CFR Part 9 implement the provisions of the Sunshine Act concerning the opening of Commission meetings to public observation and describe the procedures governing the production of agency records, information, or testimony in response to subpoenas or demands of courts or other judicial authorities in State and Federal proceedings.

5. 10 CFR Part 20

The regulations in 10 CFR Part 20, "Standards for Protection against Radiation," establish standards for protection against radiation hazards arising from licensed activities. Some material is appropriate for inclusion in the examinations administered to candidates for RO or SRO licenses.

D. Regulatory Guides, NUREG Reports, and ANSI/ANS Standards

Regulatory guides, NUREG reports, and industry standards are not requirements except as specified in Commission orders or as committed to by the facility licensee. The appropriate revisions should be consulted as referenced in the facility FSAR or approved training program. The following paragraphs summarize the latest revisions of these documents.

ANSI/ANS 15.4-1988, "American National Standard for the Selection and Training of Personnel for Research Reactors"

This standard provides criteria for the selection and training of research reactor personnel doing a variety of functions at various levels of responsibility (e.g., directors/administrators, operators, and technicians). The standard also covers the general health and disqualifying conditions applicable to license applicants and licensed personnel.

EXAMINER STANDARD 201N
PREEXAMINATION ACTIVITIES

A. Purpose

This standard describes the activities required to prepare for an operator licensing examination at a research and test reactor facility. It includes procedures for scheduling the examination, assigning an examiner(s), obtaining the facility-specific reference material, and reviewing the license applications.

B. Examination Scheduling

The chief examiner (CE) will schedule a specific examination date with the licensee's training representative. Normally, the examiner will schedule the administration of the written examination and all required operating tests during a single site visit with the operating test following the written examination.

C. Coordinating Examination Visits

Three months before the examination week, the CE should contact the facility licensee by telephone or email to reconfirm the expected number of applicants, the exact examination dates, and other arrangements for the examinations. The CE should use Attachment 1, "Research and Test Reactor Corporate Notification Letter," as a guide and discuss with the licensee the following:

- delivery of the reference materials (Attachment 1, Enclosure 1) to the CE at least 60 days before the scheduled examination date

- facility licensee responsibilities for written examinations (Attachment 1, Enclosure 2)

- applicant responsibilities for the written examination (Attachment 1, Enclosure 3)

- facility review of written examinations before administration (Attachment 1, Enclosure 4)

- the requirements of 10 CFR 55.31, "How to Apply," for submitting the license applications

The CE will normally issue a letter confirming these arrangements about 90 days before the examinations begin. This letter should be addressed to the facility licensee management representative responsible for facility operations. The exact wording of the corporate notification letter may be modified as necessary to reflect the confirmed arrangement.

Upon receipt, the CE will review the reference materials to decide if they are adequate for preparing the examination. If these materials are incomplete or inadequate, the CE will request that the facility licensee send additional reference materials.

If necessary, an examiner may review and select additional reference materials during a site orientation trip.

The CE will coordinate the examination site visit with the facility licensee and any other examiners assigned. If this will be the first visit at a particular facility for an examiner, he or she should review with the CE the need for a preexamination site visit as described in Section F of this standard.

If the examination must be rescheduled on short notice, the CE should minimize the impact on other facility examinations that have been scheduled. The same examiner should administer the entire operating test to an applicant. An examiner who failed an applicant on an operating test will not administer that applicant's retake of the operating test.

If weekend or shift work is required to administer the operating tests, the CE will coordinate with the facility licensee and, if applicable, other assigned examiners.

D. Assignment of Examiners

Depending on the number of license applicants for a given facility examination, the CE may request help to prepare and or to administer the examination. Care should be taken not to overburden any examiner. An example of a heavy week for an examiner would be testing eight initial candidates [Reactor Operator and Senior Reactor Operator (Instant)] and four Senior Reactor Operator (Upgrade) candidates, at a Complex facility. Each examiner's certification status, other examination commitments, and general availability will be considered when making assignments.

An examiner who was previously employed by a facility licensee (or one of its contractors) will not be assigned any direct responsibilities for developing or administering written examinations or operating tests at that licensee's facilities for **at least 2 years** after the examiner ended his or her employment.

If a facility examination assignment presents a conflict of interest for an examiner, that examiner will inform his branch chief/program director (BC/PD). The examiner will discuss the following areas, as applicable, with his or her BC/PD:

- the length of time the examiner worked at the facility

- the time elapsed since the examiner left the facility

- the reasons the examiner ended employment at the facility

- the nature and extent of previous relationships with persons employed at the facility

- anything that could affect the administration, performance, evaluation, or results of the examination or create the **appearance** of a conflict of interest

After reviewing the facts of the possible conflict-of-interest case to determine whether the assigned examiner can maintain his or her objectivity, the BC/PD will make the appropriate assignment determination.

E. Reviewing License Applications

The CE will review the applications according to ES-202N to decide whether the applicants meet the requirements specified in 10 CFR 55.31, 10 CFR 55.33, "Disposition of an Initial Application," and 10 CFR 55.35, "Re-Applications," for initial license candidates or the requirements of 10 CFR 55.57, "Renewal of Licenses," for license renewal.

After the CE has reviewed and approved or denied each of the applications and resolved all of the waiver requests, the CE will forward the applications to the licensing assistant responsible for research and test reactors operator licensing docket files (OLA).

The OLA will create a 55 docket folder for each new applicant and prepare letters informing the applicant of the final disposition of any application-specific waivers (refer to ES-203N). The OLA will then prepare an examination assignment sheet (Attachment 2 to this standard) as far in advance as possible. The assignment sheet will identify all assigned examiners by name and list the applicants by name, docket number, and type of examination they are to take (e.g., SRO-Upgrade (SRO-U), RO written only). All applicants listed on the assignment sheet should be administered complete examinations (written and operating) as indicated under *Examination Type* unless the NRC has granted waivers in accordance with ES-203N. A copy of the assignment sheet will be distributed to all examiners assigned and the facility project manager. The assignment sheet, which can be found in the Agencywide Documents Access and Management System (ADAMS), will be updated to reflect the examinations as they were administered.

F. Preexamination Site Visits

Occasionally, taking a trip to the facility before the scheduled examination date may be advantageous for an examiner. This will allow the examiner to familiarize himself or herself with the facility and provide an opportunity to review and validate examination materials with the facility licensee.

The branch chief (BC) will assess the need for a separate preparatory site visit. The BC should carefully weigh the costs and benefits associated with each additional trip to the facility and consider factors such as the experience of the assigned examiner(s), the number of applicants, and the complexity of the operating tests to be administered.

G. Facility Preexamination Reviews

The CE may arrange for the facility licensee to review the written examination before it is administered. The CE will conduct the examination review according to the guidelines and instructions contained in Attachment 1.

ATTACHMENTS/FORMS:

Attachment 1, Sample Corporate Notification Letter
Attachment 2, Sample Examination Assignment Sheet

ATTACHMENT 1
SAMPLE CORPORATE NOTIFICATION LETTER

[NRC Letterhead]

[Date]

[Facility Contact Name]
[Facility Contact Title]
[Street Address line 1]
[Street Address line 2]
[City, State Zip Code]

SUBJECT: CORPORATE NOTIFICATION LETTER, 50-[Number]/OL-YY-NN, [FACILITY
             NAME]

Dear [Name]:

I arranged with [facility contact or you] for the administration of operator licensing examinations at the [facility] reactor. The written and operating examinations are scheduled for the week of [date of exams].

To meet this schedule, please furnish the material listed in Enclosure 1, "Reference Material for Operator Licensing Examinations," at least 60 days before the examination date to either of the following addresses:

(U.S. Postal Service)   (Private Mail Carrier)
ATTN: [Chief Examiner]   ATTN: [Chief Examiner]
[Appropriate Mail Stop]   [Appropriate Mail Stop]
U.S. Nuclear Regulatory Commission   U.S. Nuclear Regulatory Commission
Washington, DC 20555   11555 Rockville Pike
                                               Rockville, MD 20852-2738

Enclosure 2, "Administration of Written Examinations," describes your responsibilities for conducting written examinations. Enclosure 3, "Procedures for the Administration of Written Examinations," describes applicant responsibilities during the administration of the written examination. Please ensure that all applicants are aware of these rules.

Your review of the written examination will be conducted in accordance with the procedures specified in Enclosure 4, "Facility Review of Written Examinations."

Final, signed reactor operator and senior reactor operator license applications certifying that all training has been completed must be submitted at least 14 days before the first examination dates. This will allow the chief examiner time to review the training and experience of the applicants, process the medical certifications, and process the applications. If this review cannot be completed in time to decide an applicant's eligibility, that applicant may not be permitted to sit for the examination. Therefore, it is recommended that license applications be provided as soon as possible to ensure an appropriate level of review.

The NRC has posted copies of the application forms, "Personal Qualification Statement—Licensee" (NRC Form 398) and "Certification of Medical History by Facility Licensee" (NRC Form 396) on the agency's Web site at www.nrc.gov/reading-rm/doc-collections/forms/nrc398.pdf and www.nrc.gov/reading-rm/doc-collections/forms/nrc398.pdf, respectively. These forms are in Adobe Acrobat® format.

This letter contains information collection requirements that are subject to the Paperwork Reduction Act of 1995 (44 U.S.C. 3501 et seq.). These information collections were approved by the Office of Management and Budget, approval number 3150-0018.

The burden to the public for these mandatory information collections is estimated to average 7.7 hours per response, including the time for reviewing instructions, searching existing data sources, gathering and maintaining the data needed, and completing and reviewing the information collection. Send comments regarding this burden estimate or any other aspect of these information collections, including suggestions for reducing the burden, to the Records and FOIA/Privacy Services Branch (T-5 F52), U.S. Nuclear Regulatory Commission, Washington, DC 20555-0001, or by Internet electronic mail to INFOCOLLECTS@NRC.GOV; and to the Desk Officer, Office of Information and Regulatory Affairs, NEOB-10202, (3150-0018), Office of Management and Budget, Washington, DC 20503.

The NRC may not conduct or sponsor, and a person is not required to respond to, a request for information or an information collection requirement unless the requesting document displays a currently valid OMB control number.

If you have any questions regarding the examination procedures and requirements, please contact me at (301) 415-[number], or email at [initials]@nrc.gov.

Sincerely,

[Name], Chief Examiner
[Branch or Program Name]
[Division Name]
Office of Nuclear Reactor Regulation

Docket No. 50-[number]

Enclosures:
1.   Reference Material for Reactor/Senior Operator Licensing Examinations
2.   Administration of Written Examinations
3.   Procedures for the Administration of Written Examinations
4.   Facility Review of Written Examinations

cc w/enclosures:
[Name], Reactor Supervisor, if applicable
[Name], Training Supervisor, if applicable

ENCLOSURE 1
REFERENCE MATERIAL FOR REACTOR/
SENIOR REACTOR OPERATOR LICENSING EXAMINATIONS

(1)   Training materials including all substantive written material used to prepare applicants for initial reactor operator and senior reactor operator licensing. The material should include learning objectives, if available, and details presented during lectures, rather than outlines. Training materials should be identified, bound, and indexed. Training materials should include the following:

    (a)   System descriptions including descriptions of all operationally relevant flowpaths, components, controls, and instrumentation. System training material should draw parallels to the actual procedures used for operating an applicable system.

    (b)   Learning objectives, student handouts, and lesson plans (including training manuals, facility orientation manual, system descriptions, reactor theory, thermodynamics).

    (c)   Complete and operationally useful descriptions of all safety-system interactions and secondary interactions under emergency and abnormal conditions, including consequences of anticipated operator error, maintenance error, and equipment failure.

    (d)   Training material used to clarify and strengthen understanding of emergency operating procedures.

(2)   Complete Procedure Index (including temporary procedures).

(3)   All administrative procedures as applicable to reactor operation or safety.

(4)   All integrated facility procedures, normal or general operating procedures, and procedures for experiments.

(5)   All emergency procedures, emergency instructions, and abnormal or special procedures.

(6)   Standing orders or procedures changed by reactor supervision and important orders or changes that are safety related and may supersede the regular procedures.

(7)   A list of all reactor facility surveillances, with copies of all **COMPLETED** surveillances which require the collection of data (e.g., heat balance, rod drop times).

(8)   Fuel-handling and core-loading procedures and initial core-loading procedure (if appropriate).

(9)   Any annunciator/alarm procedures, as applicable.

(10)   Radiation protection manual and radiation control manual or procedures.

(11) Emergency plan and any emergency plan implementing procedures (EPIPs).

(12) Safety analysis report, technical specifications, and interpretations, if available.

(13) System operating procedures, including experiments.

(14) Piping and instrumentation diagrams, electrical single-line diagrams, and flow diagrams.

(15) Technical data book, and/or facility curve information, as used by operators and facility precautions, limitations, and setpoints.

(16) Questions and answers specific to the facility training program, which may be used in the written or operating examinations (voluntary by facility licensee).

(17) Facility modification authorizations, which were authorized since the last revision to the safety analysis report.

(18) Additional material as requested by the examiners to develop examinations that meet the requirements of the research and test reactor examiner standards and regulations.

The above reference material should be approved, final issues and be so marked. If a facility has not finalized some of the material, the chief examiner should verify with the facility that the most complete, up-to-date material is available and that agreement has been reached with the licensee for limiting changes before the administration of the examination.

ENCLOSURE 2
ADMINISTRATION OF WRITTEN EXAMINATIONS

(1)   A single room must be provided for administration of the written examination. This room and supporting restroom facilities should be located so as to prevent contact with other facility personnel during the written examination.

(2)   Minimum spacing is necessary to ensure examination integrity. The chief examiner will determine whether the room has adequate area to support minimum spacing between examination applicants to ensure examination integrity.

(3)   The chief examiner will review any arrangements made by the facility to give the applicants lunch, coffee, or other refreshments. These arrangements shall comply with Item 1 above.

(4)   The facility licensee may provide pads of 8.5 by 11-inch lined paper in unopened packages for the applicant's use in completing the examination. The examiner will distribute these pads as needed.

(5)   Applicants may bring pens, pencils, calculators, or slide rules into the examination room. Black ink or dark pencils should be used for writing answers to questions.

(6)   The chief examiner must approve any wall charts, models, training materials, equipment, or reference material present in the examination room.

(7)   The chief examiner will give the facility staff a copy of the written examination with answer key at the beginning of the examination. The facility staff will then have 5 working days to submit formal written comments with supporting documentation regarding written examination questions and answers to the chief examiner.

ENCLOSURE 3
PROCEDURES FOR THE ADMINISTRATION OF
WRITTEN EXAMINATIONS

(1)   Verify candidate identity.

(2)   Pass out examinations and handouts. Instruct applicants not to review examination until instructed to do so.

READ THE FOLLOWING INSTRUCTIONS VERBATIM:

(1)   Cheating on the examination means an automatic denial of your application and could result in more severe penalties.

(2)   When you have completed your examination, you must sign the statement at the bottom of the cover sheet. This indicates that the work is your own and you have not received or given assistance in completing the examination.

READ THE FOLLOWING INSTRUCTIONS:

With the start of the examination, you must comply with the following rules. These rules are in effect within the examination area (DEFINE THE AREA), until the last candidate has handed in his or her examination:

(1)   Restroom trips are limited to only one applicant at a time. You must avoid all contact with anyone outside the examination room to preclude even the appearance of cheating.

(2)   Use black ink or dark pencil **only** to facilitate legible reproductions.

(3)   Print your name in the blank provided in the upper right corner of the examination cover sheet and each answer sheet.

(4)   Mark your answers on the answer sheet(s) provided.

(5)   The point value for each question is shown in brackets after the question.

(6)   If the intent of a question is unclear, ask questions of the examiner or proctor only.

(7)   There is a time limit of 1 hour per section of the written examination. For example, a one-section retake examination has a 1-hour time limit, while a normal three-section initial examination has a 3-hour time limit.

(8)   You must achieve a grade of 70 percent or greater in each category to pass the examination.

(9)   When turning in your examination, assemble the completed examination with examination questions, examination aids, answer sheets, and all scrap paper. Give the proctor your answer sheet(s) along with the signed cover sheet. Take all other material collected with you out of the examination area.

(10)   After turning in your examination, leave the examination area. If you are observed in this area while the examination is still in progress, your license may be denied or revoked.

(11)   During the examination you will be evaluated for your actions as if you were the actual watchstander. Please operate the reactor as if you were licensed, with the exception that you should announce your actions, then pause momentarily to give the operator of record time to correct you or stop you, if necessary, before you actually perform the action. In addition, the examiner will be observing that you meet all conditions of your license, (e.g., wearing corrective lenses to perform licensed duties).

ENCLOSURE 4
FACILITY REVIEW OF WRITTEN EXAMINATIONS

1.   At the option of the appropriate NRC management, the facility may review the written
      examination up to 2 weeks before its administration. This review may take place at the
      facility or an NRC office. The chief examiner will coordinate the details of the review
      with the facility. An NRC examiner will be present at all times during the review. The
      facility staff may not retain copies of the examination or any written notes.

      When using this option, the facility reviewers must sign the following statement before
      being allowed access to the examination.

a.   Preexamination Security Agreement:

I _____ acknowledge that I have acquired specialized knowledge
      [Print Name]
concerning the examination scheduled for _____ at _____
                                          [Print Date]   [Print Facility Name]
as of the date of my signature below. I agree that I will not knowingly divulge any information
concerning this examination to any unauthorized persons. I understand that I am not to
participate in any instruction involving those applicants scheduled to be administered the above
examination from this date until after the examination has been administered. I further
understand that violation of the conditions of this agreement may result in the examinations
being canceled and/or enforcement action against myself or the facility licensee by whom I am
employed or represent.

_____
              Signature/Date

In addition, the facility staff reviewers will sign the following statement after the written
examination has been administered.

b.   Postexamination Security Agreement:

I _____ did not, to the best of my knowledge, divulge any information
      [Print Name]
concerning the examinations administered during the week of _____ at _____
                                                            [Print Date]   [Print Facility Name]
or provide any instruction to those applicants who were administered the examination from the
date I entered into this security agreement until the completion of examination administration.

_____
              Signature/Date

2.   The facility staff will be provided a copy of the examination and the answer key at the
      beginning of the examination. A copy will be kept of any pen and ink changes made to
      questions during the examination administration.

If members of the facility staff did not review the examination before its administration, they will have 5 working days from the day of the written examination to submit formal comments. If the facility staff reviewed the examination before its administration, they must submit any additional comments before the examiner leaves the facility. The reactor supervisor will address the comments to the chief examiner. The NRC will consider comments submitted after the required period on a case-by-case basis. This may cause delays in grading the examinations.

3.   The facility licensee should submit comments in the following format:

   •   list the question section and number and state the comment along with a recommendation for correction (e.g., delete, two correct answers)

   •   support the comment with a reference and provide a copy of the reference if it was not included in the original reference material submitted

NOTES:

(1)   No change to the examination will be made without a reference to support the facility comment. Provide any supporting documentation not previously supplied.

(2)   Comments made without a concise facility recommendation will not be addressed.

(3)   Comments not submitted within the requested time will be included in the grading process on a case-by-case basis as determined by the NRC. Comments not submitted within the requested time will delay the examination grading process.

(4)   NRC policy is to delete a multiple-choice question that has no correct answer or more than two correct answers and to give credit for either response for questions that have two correct answers.

ATTACHMENT 2
SAMPLE EXAMINATION ASSIGNMENT SHEET

NOTE TO: (CE) [Name, Title, and Affiliation]

FROM: (OLA) [Name, Title, and Affiliation]

SUBJECT: EXAMINATION ASSIGNMENT SHEET [REPORT NO. AND FACILITY
                  NAME])

APPLICANT	DOCKET NO.	EXAMINATION TYPE

Facility and location
Facility contact
Written examination to be prepared by
Dates of examination

cc: [Project manager]
(Standard Headquarters distribution)

Examination Report No. 50-[Number]

ADAMS Accession No. ML[Number]

EXAMINER STANDARD 202N
ELIGIBILITY REQUIREMENTS AND GUIDELINES

A. Purpose

This standard provides instructions for reviewing initial and retake operator license applications. It identifies the experience, training, education, and certification requirements and guidelines that applicants should satisfy before they will be allowed to take an RO or SRO license examination.

B. Definitions

ANSI/ANS-15.4-1988, "American National Standard for the Selection and Training of Personnel for Research Reactors," defines many of the terms used in this standard. Additional clarification is provided where necessary. The following terms are relevant to the licensing process:

- **Nuclear experience:** Defined in Section 2 of ANSI/ANS-15.4-1988.

- **Academic training:** Defined in Section 2 of ANSI/ANS-15.4-1988.

- **Certification:** Defined in Section 2 of ANSI/ANS-15.4-1988.

- **Reactor operator applicant:** An unlicensed individual who is applying for an RO license.

- **Senior reactor operator-upgrade applicant:** A licensed RO who is applying for an SRO license on the same unit(s).

- **Senior reactor operator-instant applicant:** An unlicensed individual who is applying for an SRO license.

- **Senior reactor operator limited to fuel handling applicant:** An individual who is applying for an SRO license limited to fuel handling license.

C. Reviewing Initial License Applications

Subpart D, "Applications," of 10 CFR Part 55 outlines the regulatory requirements associated with the license application process. The CE should refer to them as necessary when reviewing license applications.

1. Initial License Applications

a.   All applicants must submit an NRC Form 398 and an NRC Form 396. (Computer-generated duplicates are acceptable.) An application is not complete until both forms are filled out, signed by the appropriate facility licensee personnel, and signed by the NRC.

Detailed instructions for completing NRC Form 398 are provided with the form.

b. The facility licensee senior management representative must certify that the applicant has completed the training required for the desired license level by placing a check in Item 19.b of NRC Form 398, signing the form, and submitting it to the CE at least 14 days before the examination date.

The CE will review applications against the specific RO or SRO eligibility guidelines described in Section F, process the medical certifications, evaluate any waiver requests (refer to ES-203N), and request any additional information that might be necessary.

c. If the CE decides that an application is incomplete or that the applicant does not meet the requirements in 10 CFR 55.31, he or she will note the deficiencies on the application and request additional information. If, after receipt of additional information, the CE decides that the applicant still does not meet the eligibility requirements, the BC/PD will notify the applicant in writing that the application is being denied and inform him or her of the deficiencies on which the denial is based (Attachment 1).

The CE will check the *Does Not Meet Requirements* block (bottom of NRC Form 398) and sign and date the form. Applicants who do not meet eligibility requirements will not be permitted to take a license examination.

An applicant who does not accept the proposed denial may request a review of the denial or request a hearing according to 10 CFR 2.103(b)(2). ES-502N provides further information on this topic.

2. Retake License Applications

An applicant applying after a license denial must submit new NRC Forms 398 and 396 (refer to 10 CFR 55.35). The NRC will not process an application if the applicant has a request for reconsideration or a hearing on the previous license denial outstanding (refer to ES-502N). The CE will review the application as described in Section C.1, subject to the additional conditions described in the following paragraphs.

a. If the first application was denied because the applicant failed either the written examination, operating test, or both, the applicant may file a new application 2 months after the date of the final denial. The applicant may file a third application 6 months after the date of the second denial and successive applications 2 years after the date of each subsequent denial.

b. The applicant may request a waiver of portions of the NRC examination passed on his or her **initial** examination. (Refer to ES-204 for further guidance on this type of waiver.)

c. The application will describe the extent of training since the denial.

d. Facility licensee management must certify that the applicant has received sufficient additional training at the facility and the results of that training have been objectively evaluated to ensure that the applicant is ready for reexamination, as well as safely assuming the duties and responsibilities of a licensed operator (NRC Form 398, Item 19.b).

D. Initial Licensing Medical Requirements

Subpart C, "Medical Requirements," of 10 CFR Part 55 specify the medical requirements for license applicants and licensed operators.

10 CFR Part 55 requires the applicant to submit an NRC Form 396 as part of his or her application for a license. The form should describe the guidance used by the doctor to perform the medical examination. The form lists three guides recognized by the NRC for medical examination; ANSI/ANS 3.4-1996, ANSI/ANS 3.4-1983 and ANSI/ANS 15.4-1988. If the facility uses other guidance it must list that guidance under "other". Forms stipulating "other" will be forwarded to the NRC doctor for evaluation, no licenses will be issued until the doctor completes his or her review and finds the medical examination to be adequate. The medical data in support of NRC Form 396 are normally good for 6 months from the date of the medical examination. However, if the applicant is reapplying after withdrawing a previous application or accepting a final license denial, he or she may request a waiver by checking Item 4.f.4 on NRC Form 398 and either providing an explanation in Item 17 of the form or submitting a separate letter with the application. The CE may waive the requirement for a new medical certificate (i.e., NRC Form 396) if the date of the original medical examination is within 1 year of the scheduled license examination. The CE will document the disposition of the waiver request by checking the appropriate block at the bottom of NRC Form 398 and notifying the candidate in writing (refer to ES-203N).

E. Initial License Eligibility Guidelines

The license eligibility guidelines in ANSI/ANS-15.4-1988 are summarized as follows.

1. Reactor Operator and Senior Reactor Operator—Instant Applications

a.   Certification

The facility licensee management must certify that the applicant has received sufficient training at the facility and the results of that training have been objectively evaluated to ensure the safe assumption of the duties and responsibilities of a licensed operator (NRC Form 398, Item 19.b).

b.   Training

The training shall include the following:

(1)   Sufficient design and content to ensure safe operation of the facility.

(2)   Topics identified in Sections 5.4 (RO) or 5.3 (SRO-I) of ANSI/ANS-15.4-1988.

(3)   Operation of the reactor and its systems under the supervision of licensed ROs and SROs.

(4)   Manipulation of the controls of the reactor during five significant changes in reactivity or power level (refer to 10 CFR 55.31(a)(5)). Every effort should be made to diversify the reactivity and power changes. Examples of control manipulations include reactor startups and shutdowns and large power changes.

c.   Education

The applicant should have a high school diploma or an equivalency certificate (e.g., General Education Development (GED) or home schooling certificate). Individuals who have not completed high school or do not have a home schooling certification should not be excluded. Previous job-related experience or education should also be considered. If an applicant is still attending high school, then the facility licensee should carefully review the applicant's background for selection into the training program.

2. Senior Reactor Operator—Instant Applications

The applicant should have a minimum of 3 years of nuclear experience. Two years of full-time academic training may be substituted for 2 years of experience.

3. Senior Reactor Operator—Upgrade Applications

a.   Experience

The applicant should have a minimum of 1 year of experience as an RO at the facility or meet the requirements for an SRO-Instant (SRO-I) listed above.

b.   Training

The training shall:
(1)   include the topics identified in Section 5.4 (RO) or 5.3 (SRO-I) of ANSI/ANS-15.4-1988

F. NRC Form 398

Each applicant must submit a personal qualifications statement (NRC Form 398). The form must be completely filled out per the instructions and signed by the appropriate facility licensee personnel. Those sections or items that are not applicable to operators at research and test reactors will be marked "NA."

Item 4. Type of Application

Item f. *Waiver Requested* (ES-203N discusses requests for reapplications)

Item g. *Date Passed GFE* (not applicable to research and test facilities)

Item 10. Current Position at Facility

Items a, b, c, e, and i are not directly applicable to research and test reactors. Therefore, the "other" Item j, *Other,* should be used to describe the applicant's position (e.g., facility director, chief reactor supervisor). Items g and h are applicable only to licensed personnel. An unlicensed control room operator trainee should not be listed as a control room operator in Item h but, instead, should be listed as a trainee in Item j.

Item 13. Training *(Since Last Application—See Instruction)*

This section should contain only training received specifically for the license for which the application is submitted. Research and test facilities normally do not have a training program that excludes other concurrent activities, therefore, the period of training should be identified (month and year) and condensed to the appropriate number of weeks.

For example, an applicant spends 4 months in training from June 1, 1984, through September 30, 1984, with 2 hours per week devoted to fundamentals, 2 hours per week tracing systems, 1 hour per week in the control room, and 1 hour per week in actual manipulation (two reactivity changes per manipulation). The condensed training would be from June 1984 to October 1984 and included, approximately 1 week (16 weeks × 2 hours per week) of fundamentals, 1 week of facility systems observation, 1 week of control room operations, and 32 reactivity manipulations. Numbers do not have to be precise, but should be representative.

- Items a.1, a.2, and a.3, *Classroom,* should identify training conducted in a classroom setting.

- Item b, *Simulator,* is not applicable to research and test facilities.

- Item c, *SRO Instruction,* should be used to identify SRO-level training given in preparation for an SRO-level license examination.

- Item d is not applicable to research and test facilities.

- Item f, *Other,* should be used to describe time spent in the facility in training (e.g., time spent tracing systems, time in the control room observing or operating).

- ANSI/ANS-15.4-1988 contains additional guidance on training criteria.

Item 15. Experience

Experience must be current up to the date of application. Applicable experience gained at other research and test facilities and/or nuclear power plants, as well as any military experience, should be included. In all cases, the applicant should briefly and fully describe his or her experience in Item 15.

Item 16. For Renewals Only (See ES-204)

Item 19. Signatures

Item a. The applicant's signature must appear on the application.

Item c: The training coordinator and highest level of facility management responsible for facility operations must sign the application. This is normally the facility director or equivalent position; higher authority is not needed. If the facility director is also the training coordinator, then he or she must sign both places.

G. Maintaining Medical Standards for Licensees

1. Temporary Disability

When an operator is temporarily unable to meet medical standards, the facility licensee may administratively classify that operator's license as "inactive" until he or she is again certified to meet all medical standards by the facility licensee. The facility licensee need not notify the NRC nor request a conditional license for the temporary disability provided the operator is administratively prevented from performing licensed duties during his or her temporary disability.

If the disability extends beyond the date of license expiration, the licensee may apply for timely license renewal according to 10 CFR 55.55(a) and 10 CFR 55.57(a). The facility licensee should document the nature of the licensee's temporary disability on the medical certificate and submit a revised certificate to the NRC after the physician determines that the licensee meets the requirements of 10 CFR 55.33(a)(1). The NRC will not renew the operator's license until it finds that all of the conditions specified in 10 CFR 55.57(b) are satisfied.

2. Permanent Disability

If the facility licensee determines that an operator's medical condition is permanently disqualifying according to Section 7.2 of ANSI/ANS 15.4-1988, the facility licensee shall notify the NRC within 30 days of learning of the diagnosis (see 10 CFR 50.74 and 10 CFR 55.25).

ATTACHMENTS/FORMS:

Attachment 1, Sample Initial Application Denial

ATTACHMENT 1
SAMPLE INITIAL APPLICATION DENIAL

[NRC Letterhead]

[Date]

[Applicant Name]
[Street Address]
[City, State Zip Code]

Dear [Name]:

This letter is to inform you that your application for a [reactor operator, senior reactor operator] license submitted for the [facility name] is hereby denied.

[Branch to discuss deficiencies and which part of 10 CFR 55.31, ES-202N, or ANSI/ANS-15.4-1988 was involved.] When you have met the requirements of Title 10, Section 55.31, "How to Apply," of the *Code of Federal Regulations* (10 CFR 55.1), you may submit another application.

If you do not accept this denial, you may, within 20 days of the date of this letter, take one of the following actions:

(1)   You may request the NRC to reconsider the denial of your application by writing to the Director, Division of Regulatory Improvement Programs, Office of Nuclear Reactor Regulation, U.S. Nuclear Regulatory Commission, Washington, DC 20555. Your request must include specific reasons for your belief that your application was improperly denied. If the NRC determines that the denial of your application remains appropriate, you still have the right to a hearing pursuant to 10 CFR 2.103(b)(2).

(2)   If you do not request reconsideration, you may request a hearing in accordance with 10 CFR 2.103(b)(2). Submit your request, in writing, to the Secretary of the Commission, U.S. Nuclear Regulatory Commission, Washington, DC 20555, with a copy to the Assistant General Counsel for Hearings, Office of the General Counsel, at the same address.

If you have any questions, please contact [name] at [telephone number] or via email at [initials]@nrc.gov.

                              Sincerely,

                              [Branch Chief or Program Director with Program
                              Responsibility]

Docket No. 55-[number]

cc: [Facility representative who signed the applicant's NRC Form 398]

CERTIFIED MAIL—RETURN RECEIPT REQUESTED

ELIGIBILITY REQUIREMENTS AND GUIDELINES   ES-202N

EXAMINER STANDARD 203N
PROCESSING OPERATOR LICENSING WAIVERS

A. Purpose

This standard provides guidance for processing waiver requests from RO and SRO license applicants.

B. Background

To speed the processing of an applicant's waiver request(s), the NRC management has delegated the authority to grant waivers of certain operator licensing requirements to the CE. The types of waiver requests identified in Section D.1 of this standard may be granted by the CE without first obtaining BC/PD approval.

C. General Guidelines

1. Submittal of Waiver Requests

The applicant may request a waiver of a license requirement by checking the appropriate block in Item 4.f on NRC Form 398 providing a justification in Item 17.

2. Evaluation of Waiver Requests

The CE will evaluate waiver requests on a case-by-case basis. If there is insufficient information upon which to base a decision, the CE will request additional information from the facility licensee according to ES-202N.

Waiver requests may be denied if the evaluation and judgment of the staff so warrant. The staff will document the disposition of every waiver request, whether granted or denied, by completing the *For NRC Use* block on the applicant's application.

3. Notification of Applicants

The staff will notify the applicant in writing of the disposition of the request(s) when the decision to grant or deny their waiver is made. If time is too short to notify the applicant in writing before the examination date, the CE will notify the facility licensee's training representative by telephone, with a memorandum to the applicant's docket file.

If the applicant is determined to be ineligible to take the license examination, the staff will issue a denial letter according to ES-202N.

D. Waivers

1. Routine

a.   If an applicant fails any section of the written examination, but has a passing grade overall for the examination, the CE may waive those examination sections passed. For candidates who have failing grades for the examination overall, the CE may grant a waiver for any sections for which the candidate scored greater than 80 percent. These waivers are only applicable for the first retake examination, and that examination must be applied for and taken within 1 year of the date of the examination that the applicant failed.

b. The medical data in support of NRC Form 396 are good for 6 months from the date of the medical examination for a person applying for an RO or an SRO-I license. For reapplications following a license denial or withdrawal of an application, waivers extending the 6-month period may be granted if the date of the original medical examination is within 1 year of the scheduled reexamination. For renewal and SRO-U applicants, the medical examination documented on NRC Form 396 is good for 2 years from the date of the medical examination.

c. The NRC staff will consider examination waivers for previously licensed operators not currently holding a license depending on the justification provided by the facility licensee pursuant to 10 CFR 55.47, "Waiver of Examination and Test Requirements."

2. Nonroutine Waivers

For all other waiver requests, the CE will make the appropriate recommendations to the NRC management. The NRC management will grant or deny the waiver request. NRC management concurrence is not required for a waiver decision made by the CE in accordance with Section D.1 above.

EXAMINER STANDARD 204N
REVIEWING LICENSE RENEWAL APPLICATIONS

A. Purpose

This standard provides instructions for reviewing renewal license applications.

B. Timely Renewal

All operators must comply with the requirements of 10 CFR 55.57(a) to renew his or her license.

(1) According to 10 CFR 55.55(b), "If an individual licensee files an application for renewal or upgrade of an existing license on Form NRC-398 at least 30 days before the expiration of the existing license, it does not expire until disposition of the application for renewal or for an upgraded license has been finally determined by the Commission."

(2) If an individual licensee is waiting for reexamination after failing an NRC-conducted requalification examination, his or her license will be extended under the provisions of 10 CFR 55.55(b) until the NRC makes a renewal decision.

(3) If the application is received less than 30 days before the date of license expiration and too late for processing, the license will expire on the expiration date. A new license may be issued when processing of the application is completed.

C. Reviewing License Applications

(1) The operator must submit NRC Form 398 and NRC Form 396 (computer-generated duplicates are acceptable). The application is not complete until both forms are filled out, signed by the appropriate personnel, and signed by the NRC.

Detailed instructions for completing NRC Form 398 are provided with the form. Section D of this standard provides additional information.

(2) 10 CFR 55.21, "Medical Examination," requires the facility licensee to submit an NRC Form 396 certifying that a physician has performed a medical examination. The form must show that the physician performed the examination within the previous 2 years, and stipulate the guidance used to perform the examination (e.g., ANSI/ANS 3.4-1983, ANSI/ANS 3.4-1996, or ANSI/ANS 15.4-1988).

(3) The CE will review the application, process the medical certification, evaluate any waiver requests, and request any additional information that might be necessary.

(4) The CE will approve the application and renew the license if he or she finds that the conditions in 10 CFR 55.57(b) are satisfied.

(5) Upon deciding that the application is incomplete, the CE will note the deficiencies, request the facility licensee to supply the additional information within 20 days, check the *Does Not Meet Requirements* block at the bottom of Form 398, and sign and date the form. The operator license will not be renewed until the operator meets all requirements.

(6)   If, after reviewing the additional information, the CE decides that the operator still does not meet the requirements for license renewal, further action will be taken according to the instructions in Section E of this standard.

D. NRC FORM 398

Each operator must submit a personal qualifications statement (NRC Form 398). The form must be completely filled out per the instructions and signed by the appropriate personnel.

Item 4. Type of Application

Item f. *Waiver Requested* (ES-203N discusses waiver requests)

Item g. *Date Passed GFE* (not applicable to research and test reactor facilities)

Item 10. Current Position at Facility

Items a, b, and e are not applicable to research and test reactor facilities.

Item j, *Other*, should be used to describe facility staff positions (e.g., facility director, chief reactor supervisor).

Item 13. Training *(Since Last Application—See Instruction)*

Item e, *Requalification*—The use of "continuous" or a similar type of entry to describe the number of weeks in this block is not sufficient. The actual number of weeks (condensed into one time period) spent in requalification must be listed as described in the example below.

Each operator should enter the number of weeks spent on requalification training during the 6-year license period. For example, if the operator spent 10 months of each year in training with 2 hours per week devoted to study, the condensed training would be 6 years × 42 weeks/year × 2 hours/week = 512 hours. Divided by 40, this is approximately 13 weeks. Numbers do not need to be precise, but should be representative.

ANSI/ANS-15.4-1988 contains additional guidance on training criteria.

Item 14. Significant Control Manipulations

Not required to be documented for license renewal.

Item 15. Experience

Not required to be documented for license renewal.

Item 16. For Renewals Only

Item a. Enter approximate hours operated facility.

Item b. Include the date and result of most recent facility requalification examination.

Item 19. Signatures

Item a. The operator's signature must appear on the application.

Item c. Check the box, then have the facility licensee training coordinator and highest level of management responsible for facility licensee operations sign the application. This is normally the facility director or equivalent position; higher authority is not needed. If the facility director is also the training coordinator then he or she must sign in both places.

E. Procedure for Denying an Application for License Renewal

If, after receipt of additional information, the CE decides that the applicant still does not meet the eligibility requirements, the responsible BC/PD will notify the applicant in writing that the application is being denied and inform him or her of the deficiencies upon which the denial is based (Attachment 1).

Within 20 days of the date of the letter of notification of proposed denial, the individual licensee may exercise one of the following options:

(1)   Submit a written request for the NRC to review the application. Such requests should be sent to either of the following addresses:

(U.S. Postal Service)   (Private Mail Carrier)
Appropriate Division Director   Appropriate Division Director
U.S. Nuclear Regulatory Commission   U.S. Nuclear Regulatory Commission
Washington, DC 20555   11555 Rockville Pike
                                                          Rockville, MD 20852-2738

The individual licensee should include the reasons for the review request and supporting documentation as applicable. Note 5 in Section G of this standard provides additional information on the requirements for supporting documentation.

(2)   Submit a written request for a hearing pursuant to 10 CFR 2.103(b)(2). A hearing request is required to be submitted to either of the following addresses:

(U.S. Postal Service)   (Private Mail Carrier)
Secretary of the Commission   Secretary of the Commission
U.S. Nuclear Regulatory Commission   U.S. Nuclear Regulatory Commission
Washington, DC 20555   11555 Rockville Pike
                                                          Rockville, MD 20852-2738

        Send a copy to the Assistant General Counsel for Hearings, Office of General Counsel, at the same address.

If the individual licensee exercises option (1) and the staff sustains its denial of the application, the applicable division director will so inform the individual licensee using Attachment 2. The individual licensee again has the option to request a hearing pursuant to 10 CFR 2.103(b)(2).

F. Procedure for Overturning Renewal Denials

If, during a hearing or an informal review, the staff reverses its decision regarding the failure of a requalification examination or application denial, the staff will take one of the following three actions, as appropriate:

(1)   reinstate the license

(2)   allow the individual licensee to renew the license pursuant to 10 CFR 55.57, if all other requirements are satisfied

(3)   allow the individual licensee to perform licensed duties when he or she has successfully completed the facility's requalification program and the provisions of 10 CFR 55.53(e) or (f)

If, upon conducting a hearing, the staff reverses its denial of an individual licensee's renewal application, the individual licensee will be eligible for license renewal pursuant to 10 CFR 55.57 if all other requirements that were not at issue in the hearing are satisfied.

G. Notes

(1)   Letters informing an individual licensee of a proposed denial or examination failure must be signed by NRC management (BC/PD or higher). In case of an appeal, a copy of the Division's correspondence will be distributed to the BC/PD for tracking purposes.

(2)   The NRC will provide copies of all correspondence related to this process to the facility licensee's representative authorized to sign the renewal application.

(3)   The NRC should send all correspondence related to this process to the individual licensee by certified mail with return receipt requested.

(4)   Asking the facility licensee to reassess the need for the individual licensee's license is inappropriate while conducting an informal review or hearing.

(5)   Requests for informal reviews by the NRC must (a) list the items for which additional review is being requested and (b) include documentation supporting the contentions made by the individual licensee. The package containing the supporting documentation for the review and the review request must be mailed or delivered to the applicable division director within 20 days of the date of the failure or denial notification. The division staff should complete the review within 45 days of receiving the package. The staff will review requests using the guidance in ES-502N.

ATTACHMENTS/FORMS:

Attachment 1, Sample Renewal Application Denial Letter (Branch, Program Level)
Attachment 2, Sample Renewal Application Denial Letter (Division Level)

ATTACHMENT 1
SAMPLE RENEWAL APPLICATION DENIAL LETTER

[NRC Letterhead]

[Date]

[Operator's Name]
[Street Address]
[City, State Zip Code]

Dear [Name]:

This letter is to inform you that your renewal application for a [reactor operator, senior reactor operator] license submitted for the [facility name] is hereby denied.

[Branch to discuss deficiencies and which part of 10 CFR 55.31, ES-202N, or ANSI/ANS-15.4-1988 was involved.] When you have met the requirements of Title 10, Section 55.31, "How to Apply," of the *Code of Federal Regulations* (10 CFR 55.31), you may submit another application.

If you do not accept this denial, you may, within 20 days of the date of this letter, take one of the following actions:

1.   You may request the NRC to reconsider the denial of your application by writing to either of the following addresses:

(U.S. Postal Service)   (Private Mail Carrier)
[Applicable Division Director]   [Applicable Division Director]
U.S. Nuclear Regulatory Commission   U.S. Nuclear Regulatory Commission
Washington, DC 20555,   11555 Rockville Pike
                                               Rockville, MD 20852-2738

Your request must include specific reasons for your belief that your application was improperly denied. If the NRC determines that the denial of your application remains appropriate, you still have the right to a hearing pursuant to 10 CFR 2.103(b)(2).

2.   If you do not request reconsideration, you may request a hearing according to 10 CFR 2.103(b)(2). Submit your request, in writing, to either of the following addresses:

(U.S. Postal Service)   (Private Mail Carrier)
Secretary of the Commission   Secretary of the Commission
U.S. Nuclear Regulatory Commission   U.S. Nuclear Regulatory Commission
Washington, DC 20555,   11555 Rockville Pike
                                               Rockville, MD 20852-2738

Send a copy to the Assistant General Counsel for Hearings, Office of the General Counsel, at the same address.

If you have any questions, please contact [name] at [telephone number] or via email at [initials]@nrc.gov.

Sincerely,

[Branch Chief or Above]
[Appropriate Branch or Directorate]
[Appropriate Division]
Office of Nuclear Reactor Regulation

Docket No. 55-[number]

cc: [Facility representative who signed the operator's NRC Form 398]

CERTIFIED MAIL—RETURN RECEIPT REQUESTED

ATTACHMENT 2
SAMPLE RENEWAL APPLICATION DENIAL LETTER

[NRC Letterhead]

[Date]

[Operator's Name]
[Street Address]
[City, State Zip Code]

Dear [Name]:

This letter is to inform you that my staff has reviewed the proposed denial of your renewal application for a [reactor operator, senior reactor operator] license.

[Division staff to discuss deficiencies and which part of 10 CFR 55.31, ES-204N, or ANSI/ANS-15.4-1988 was involved.] When you have met the requirements of Title 10, Section 55.57, "Renewal of Licenses," of the *Code of Federal Regulations* (10 CFR 55.57), you may submit another application.

If you do not accept this denial, you may, within 20 days of the date of this letter, request a hearing pursuant to 10 CFR 2.103(b)(2). Submit your request, in writing, to one of the following addresses:

(U.S. Postal Service)	(Private Mail Carrier)
Secretary of the Commission	Secretary of the Commission
U.S. Nuclear Regulatory Commission	U.S. Nuclear Regulatory Commission
Washington, DC 20555	11555 Rockville Pike
	Rockville, MD 20852-2738

Send a copy to the Assistant General Counsel for Hearings, Office of the General Counsel, at the same address.

If you have any questions, please contact [name] at [telephone number].

Sincerely,

[Name], Director
[Appropriate Division]
Office of Nuclear Reactor Regulation

Docket No. 55-[number]

cc: [Facility representative who signed the operator's NRC Form 398]

CERTIFIED MAIL—RETURN RECEIPT REQUESTED

EXAMINER STANDARD 301N
PREPARATION OF OPERATING TESTS

A. Purpose

As required by 10 CFR Part 55, all applicants for an RO or SRO license must take an operating test, unless it has been waived by 10 CFR 55.47 (refer to ES-203N). The specific content of the examination will depend upon the type of license and the type of facility for which the applicant has applied. This standard describes the procedure for developing operating tests according to the requirements of 10 CFR 55.45, "Operating Tests."

B. Definitions

- **Applicant:** An individual who has submitted an NRC Form 398 in pursuit of an operator license. For purposes of this and other examiner standards, it is synonymous with "candidate." See the individual license level definitions in ES-202 B.

- **Section:** A major subdivision of related subjects on the operating test.

- **Operating test:** The portion of the operator licensing examination based upon direct interaction between an examiner and an applicant. It tests the applicant's knowledge of the design and operation of the reactor.

- **Scenario:** An integrated group of events comprising a set of facility malfunctions and evolutions performed or discussed using the reactor.

C. Regulatory Basis

The operating test requires the applicant to show an understanding of and the ability to perform the actions necessary to accomplish a representative sample from among the following 13 items identified in 10 CFR 55.45(a):

(1) Perform prestartup procedures for the facility, including operation of those controls associated with facility equipment that could affect reactivity.

(2) Manipulate console controls as required to operate the facility between shutdown and designated power levels.

(3) Identify annunciators, conditions, and indicating signals and perform remedial action, where appropriate.

(4) Identify the instrumentation systems and the significance of facility instrument readings.

(5) Observe and safely control the operating behavior characteristics of the facility.

(6) Perform control manipulations required to obtain desired operating results during normal, abnormal, and emergency situations.

(7) Safely operate the facility's heat removal systems, including primary coolant, emergency coolant, and decay heat removal systems and identify the relation of the proper operation of these systems to the operation of the facility.

(8)   Safely operate the facility's auxiliary and emergency systems, including operation of those controls associated with facility equipment affecting reactivity or the release of radioactive materials to the environment.

(9)   Display or describe the use and function of the facility's radiation monitoring systems, including fixed radiation monitors and alarms, portable survey instruments, and personnel monitoring equipment.

(10)   Display knowledge of significant radiation hazards, including levels more than those authorized, and the ability to perform procedures to reduce excessive levels of radiation and to guard against personnel exposure.

(11)   Display knowledge of the emergency plan, including, as appropriate, the RO or SRO responsibility to decide whether the plan should be executed and the duties under the plan assigned.

(12)   Display the knowledge and ability (K/A), as appropriate to the assigned position, to assume the responsibilities associated with the safe operation of the facility.

(13)   Display the ability to function within the control room team as appropriate to the assigned position, assuring that facility procedures are followed and that the license limitations and amendments are not violated.

D. Level of Examination—Facility Type

Research and test reactors vary widely in their complexity. To take this into account, examinations are developed using a graded approach. The facilities have been classified into three levels of complexity—Complex, Moderate, and Simple. Complex facilities are those licensed to operator at 500 kilowatts or greater. Simple facilities are AGN-200 series reactors, and Moderate facilities are all other research and test reactors.

Attachment 2 to this standard contains a table listing facilities by type. Please note that this table may be modified to move facilities within complexity types without the need to revise the standard.

The following guidelines may be helpful in differentiating between operating test levels for the three levels of facility complexity.

All operating tests will contain three categories—Category A (Administrative Topics), Category B (Facility Walkthrough), and Category C (Integrated Facility Operations). However, the requirement for the minimum number of systems tested in Category C is facility-level dependent.

The following definitions describe the levels of complexity:

*   **Complex:** A minimum of nine systems tested covering as many of the six areas as applicable.

*   **Moderate:** A minimum of six systems tested covering as many of the six areas as applicable.

- **Simple:** A minimum of three systems tested covering as many of the six areas as applicable.

E. Level of Examination—License Type

The depth and scope of operating tests are also dictated by the license for which the applicant applies. In addition to the knowledge, skills, and abilities (KSAs) monitored for RO applicants, the NRC examiner must evaluate SRO applicants on additional KSAs, such as their ability to evaluate facility conditions during abnormal and emergency conditions and to direct others, both reactor staff and outside support agencies, in recovery operations. All SRO applicants will be examined for the highest on-shift position (e.g., shift supervisor).

The examiner's review of the reference material may supply some guidance in developing an examination that is in concert with the assignment of duties and responsibilities at that facility.

The following guidelines may be used to help differentiate between RO and SRO operating tests:

(1)   SROs must have a broader, more thorough knowledge of facility administrative controls, including limitations imposed by regulations and limitations set forth in the technical specifications, along with their bases.

(2)   SROs are often assigned comprehensive actions during emergencies and abnormal conditions and should display knowledge of these assignments. If the facility's procedures allow the SRO to be on call, the RO should display this knowledge.

(3)   SROs are often assigned responsibilities for auxiliary systems outside the control room that are not normally operated by licensed operators. A common example is the waste disposal and handling system for which the licensed operator's responsibility ends when the fluid passes the last instrument that has console display. However the SRO is often responsible for ensuring that facility limits for maximum permissible concentration, effluent release rates, and other aspects are not exceeded.

The SRO-I operating test is the most comprehensive because the applicant must be evaluated for both RO and SRO levels of responsibility. The examiner must be assured that the applicant has the necessary skills and abilities as an RO and has the required knowledge and supervisory capabilities to function as an SRO.

F. Source Material

The content of the operating test will be based on licensed RO or SRO duties as described in the safety analysis report, system description manuals, operating procedures, the facility license and license amendments, licensee event reports, and other materials received from the facility licensee. (NOTE: This is the material requested from the facility in the preexamination letter described in ES-201N.) In addition, the examination may contain RO and SRO duties described in 10 CFR Part 50 and 10 CFR Part 55.

G. Requirements

The operating test is divided into three major categories—Category A (Administrative Topics), Category B (Facility Walkthrough), and Category C (Integrated Facility Operations).

When preparing operating tests, the examiner should pre-script questions to the maximum extent possible. There will be some questions for which the answer will not be known until the examiner is on site (especially for modified systems and procedures). In addition, any questions asked to followup on perceived applicant weaknesses must be documented for subsequent review and grading purposes.

1. Category A

Category A covers topics associated with the administrative control of the facility which are divided into three groups. The examiner should cover all subjects and all groups as applicable to the facility. The preparation and administration of Category A questions will be basically the same at all three types of facility.

In developing Category A questions, the examiner may use previously developed questions from either an NRC or facility examination bank or may develop new open-reference questions. In either case, the questions asked should be pre-scripted to the maximum extent possible. Any questions asked to followup on perceived applicant weaknesses must be documented for subsequent review and grading purposes.

The following descriptions should be used as guidelines for developing or selecting questions to ascertain/confirm minimal competency within each subcategory:

a. Subcategory A.1

Subcategory A.1 evaluates the applicant's knowledge of the administrative requirements associated with the operation of the facility. For example, information dissemination questions may be asked within the framework of conducting a shift turnover or integrated into other discussions as they apply throughout the test.

b. Subcategory A.2

Subcategory A.2 addresses radiation protection, a subject area in which there is significant deviation between the knowledge expected of an RO and that expected of an SRO. RO responsibilities generally entail knowledge associated with radiation worker responsibilities and operation of facility systems associated with liquid and gaseous waste releases. SROs, however, are also involved in the approval of release permits and should be aware of the requirements associated with those releases and their potential effect on the health and safety of the public. A discussion or simulated performance of a planned release may be used when examining these topics.

If possible, the examiner should enter a radiologically controlled area during the facility walkthrough. This offers an excellent opportunity to discuss most of the radiological control subjects in Subcategory A.3. A task with followup questions is also an appropriate method for performing an evaluation of this subcategory.

c. Subcategory A.3

In Subcategory A.3, there are significant differences between the knowledge required of ROs and that required of SROs for all three topics—the Emergency Plan, Fuel Movements, and the Security Plan. For the emergency plan, ROs typically only have to be able to implement the facility EPIPs. A basic familiarity with the plan and his or her responsibilities is appropriate for the RO applicant.

SROs must display knowledge based upon their responsibility to direct and manage the implementation of the EPIPs during the initial phases of an emergency. SROs must be familiar with event classification procedures and communication requirements and methods and have a more detailed overall understanding of the EPIPs. These requirements also apply to an RO if the facility's procedures assign the RO as acting emergency director in the absence of the SRO.

Fuel handling should be covered in the fuel-handling areas of the facility. The knowledge and skills should be appropriate to the facility-specific requirements for the applicant's license level. Examiner guidance is best found in the facility's procedures associated with fuel-handling operations.

The RO should be aware of his or her duties in the control room during fuel handling (i.e., alarms from the fuel-handling area, communication with the fuel storage facility, systems operated from the control room in support of fueling operations, and supporting instrumentation). For the SRO, this subject should cover information such as the delivery of new fuel, moving new/spent fuel, storage of new/spent fuel, design of the fuel-handling area, tools used, and casualty operations.

An RO applicant may be evaluated on security by observing his or her behavior during the examination. An SRO applicant should be questioned on applicable aspects of the facility's security plan.

2. Category B

This category tests the applicant's knowledge of system design and operation. The examiner will evaluate the applicant's ability to perform tasks and to answer questions on specific systems.

Attachment 1, "Systems for Operating Tests," lists examples of systems typically found at research and test reactor facilities. Note that Attachment 1 may not be all inclusive. The examiner should select systems from this list as appropriate for the specific facility. If an examination is administered over the course of 2 or more days, the examiner should vary coverage of systems and subjects across test administrations to enhance test integrity.

For Complex facilities, the examiner should evaluate RO and SRO-I applicants on at least eight systems from all applicable categories listed in Attachment 1. For Moderate facilities, the examiner should evaluate RO and SRO-I applicants on at least six systems from all applicable categories listed in Attachment 1.

For SRO-U applicants from all facilities and RO and SRO-I applicants from Simple facilities, the examiner should evaluate each applicant on at least three systems from at least three different categories listed in Attachment 1.

For all SRO applicants, the examiner will include questions and/or one task evaluating the applicant's ability to carry out actions required during an emergency or abnormal conditions.

Tasks and pre-scripted questions may be selected from existing facility examination banks, if available. The examiner may choose to annotate procedures with clarifying comments on how to execute particular steps, as well as identifying critical steps. To evaluate a subject area satisfactorily, the examiner will ask enough questions to determine the applicant's knowledge.

When using the facility's bank, the examiner may use no more than 10 percent of the questions associated with the particular system being evaluated. If the examiner determines that additional questions must be asked to clarify the observed performance, the question and the response must be fully documented to allow for postexamination review.

3. Category C

Scenarios must be prepared in advance to ensure a proper balance of events and sufficient basis to evaluate operator competency in all required K/As. The examiner must ensure that the scenarios adequately cover the following integrated facility operations test requirements:

(1)   Each scenario must require the applicant to operate during normal and reactivity evolutions, instrument failures, component failures, and major facility transients. The table on the following page specifies the minimum requirements for the type and number of events for a scenario set.

(2)   In addition to the events, there is a requirement to examine individual competencies for the RO and SRO license levels. Form ES-303N-1 describes these competencies in detail. For clarity, these two major requirements are discussed separately, although they are used integrally during scenario development, administration, and grading.

a. Transients and Events

A scenario is a set of events designed to cover a combination of the six different types of facility evolutions, equipment failures, and facility transients discussed below (meeting the requirements of the table at the end of this standard).

- **Normal evolutions:** Examples include system lineups, prestartup checkouts, shifting rod control mode between automatic and manual, and critical checks (e.g., excess reactivity measurements).

- **Reactivity evolutions:** Examples include reactor startup, power maneuvering, experiment manipulation, and pulsing operations.

- **Instrument failures:** These include failures of nuclear, control, process, and radiation detection instrumentation.

- **Component failures:** These include failures of significant reactor equipment or components that cause significant system response and require operator action to correct. Examples of component failures are control rod failure (stuck or dropped), pump failure, and piping failure.

- **Major transients:** These include facility transients that lead to an automatic protective action. Examples are reactor trip/scram or engineered safety system actuation.

- **Major transients (with entry into emergency plan):** These include transients that lead to automatic protective actions such as a reactor trip/scram or an engineered safety system actuation with subsequent entry implementation of the facility emergency procedure and/or emergency implementing procedures.

As required by 10 CFR 55.45a(2), all RO and SRO-I applicants must perform a reactor startup. The startup should include performance of prestartup checks (which provide a good opportunity for system-related discussions), startup of the facility to a critical condition, ascension to a typically maintained power level, and placing control of the reactor in automatic (if available). After completing the startup, control of the reactor may be turned over to a licensed operator. In the extraordinary situation in which a system failure occurs after the examiners arrive on site that precludes reactor startup, examinations should be rescheduled to a period when the reactor is available for this evolution.

RO applicants are required to operate the reactor during at least one normal evolution and one reactivity transient and to react to at least one instrument failure, one component failure, and one major facility transient (actual or simulated).

Scenarios for SRO-I applicants will include all transients types tested for an RO applicant plus an additional major facility transient tailored to SRO supervisory skills and knowledge of the facility's emergency plan. SRO-I applicants will be evaluated for both operational and supervisory K/As.

SRO-U applicants will be given credit for having an NRC RO license and normally will not be required to manipulate the controls or be evaluated during a reactivity evolution. However, if the examiner detects weaknesses during the test, additional scenarios, including a startup, may be used to observe and evaluate the applicant's manipulative skills.

Also, the SRO-U applicant will be tested on only one major transient tailored to SRO supervisory skills and knowledge of the facility's emergency plan. The SRO-U will be evaluated primarily for supervisory K/As.

These minimum requirements for scenario events are intended to ensure that a range of events and evolutions is exercised in each reactor facility test. Scenarios should be developed such that a variety of systems are affected within each type of event (i.e., normal evolutions, instrument failures, component failures, and major facility transients). The severity of events, and the demands they place on the applicants, should be balanced to allow each applicant to show competency across a range of conditions.

Applicant	RO	SRO-I	SRO-U
Normal Evolution	1†	1†	0
Reactivity Evolution	1†	1†	0
Instrument Failure	1‡	1‡‡	1·
Component Failure	1‡	1‡‡	1·
Major Transient	1‡	1‡‡	0
Major Transient (Implementation of Emergency Plan)	0	1‡‡	1·

† Normal and Reactivity evolutions may be combined for RO and SRO-I candidates.

‡ Either Instrument or Component failures may be combined with Major Transient for RO candidates.

‡‡ Either Instrument Failure or Component Failure may be combined with either Major Transient or Major Transient (Implementation of Emergency Plan) for SRO-I candidates, however, the Major Transient and Major Transient (Implementation of Emergency Plan) may not be combined.

·   Either Instrument Failure or Component Failure may be combined with Major Transient or Major Transient (Implementation of Emergency Plan) for SRO-U candidates.

b. Scenario Development Guidelines

To achieve maximum benefit using the reactor facility for evaluation, the following guidelines are recommended in preparing scenarios:

- Normal evolutions can be used as a backdrop upon which to stage the emergency or abnormal situations. For example, an examiner may provide verbal cues that the reactor period indication is not responding to a normal power change.

- Selected short surveillances may be used to examine panel dexterity (e.g., exercising safety rods) and should be combined with other activities, such as a reactor startup.

- The interactions of systems and components can be used to evaluate an operator's knowledge by having one failure cause or exacerbate another. Within each scenario, the events should be selected such that successive equipment failures lead to a gradual and logical deterioration in facility status; a series of unrelated and isolated events should be avoided. Avoid having one failure fully mask the symptoms of another because this can cause a scenario set to be deficient in covering the required types of evolutions. The scenarios should incorporate facility-specific and generic industry operating experience.

ATTACHMENTS/FORMS:

Form ES-301N-1, Research and Test Reactor Operator Licensing Individual Examination Report
Attachment 1, Topics for Operating Tests
Attachment 2, Facility Listing by Type

NRC FORM ES-301N-1

RESEARCH AND TEST REACTOR
OPERATOR LICENSING INDIVIDUAL EXAMINATION REPORT

CANDIDATE NAME:	CANDIDATE DOCKET NUMBER:

FACILITY NAME:	FACILITY COMPLEXITY:

INITIAL EXAMINATION		RETAKE EXAMINATION	
REACTOR OPERATOR		REACTOR OPERATOR	
SENIOR OPERATOR (INSTANT)		SENIOR OPERATOR (INSTANT)	
SENIOR OPERATOR (UPGRADE)		SENIOR OPERATOR (UPGRADE)	

WRITTEN EXAMINATION SUMMARY

PREPARED BY:	DATE ADMINISTERED:	GRADED BY:
P	01/ /2007	P

CATEGORY A SCORE	CATEGORY B SCORE	CATEGORY C SCORE	OVERALL SCORE
%	%	%	%

OPERATING TEST SUMMARY

ADMINISTERED BY:			DATE:

CATEGORY A SCORE	CATEGORY B SCORE	CATEGORY C SCORE	OVERALL SCORE

GRADING RESULTS

	PASS	FAIL	WAIVE		
WRITTEN EXAMINATION				GRADER:	DATE:
OPERATING TEST				GRADER:	DATE:
POST EXAMINATION PEER REVIEWER				GRADER:	DATE:

LICENSE RECOMMENDATION

	ISSUE LICENSE	CHIEF EXAMINER:	DATE:
	DENY LICENSE		

NRC FORM ES-301N-1		Page 2 of 6

RESEARCH AND TEST REACTOR
OPERATOR LICENSING INDIVIDUAL EXAMINATION REPORT

FACILITY NAME:	CANDIDATE DOCKET NUMBER:

Operating Test

Category A: ADMINISTRATIVE TOPICS

SUBJECT AREA	Individual Evaluation
1. Support and Conduct of Operations	
a. Modifications to Procedures and Equipment (e.g., 10 CFR 50.59 and facility procedures/forms)	
b. Surveillance Testing and Corrective Maintenance Requirements (e.g., ability to isolate equipment and components (tagouts, equipment lockouts), surveillance procedures and official form location)	
c. Information Dissemination (e.g., night orders, shift turnover logs, status boards)	
d. Startup Requirements (e.g., prestartup procedures/forms, technical specifications, required equipment)	
e. Operational Limits and Requirements (e.g., SL, LSSS, LCO, shift turnover process)	
2. Radiation Health Physics	
a. Radiation Sources and Hazards (e.g., N¹⁶, Ar⁴¹, H³, Na²⁴, open beam ports, sample removal, pneumatic tubes)	
b. Portable Radiation Monitoring Equipment (e.g., GM tubes, ion chambers, scintillation detectors, pocket dosimeters)	
c. Exposure Limits and Control (e.g., 10 CFR Part 20 limits, local facility limits, deep dose vs. shallow dose, ALI)	
d. Radiation Work Permits/Release of Radioactive Materials (including waste).	
3. Implementation of NRC-Required Procedures	
a. Security (at the proper level (RO/SRO/management))	
b. Fuel Handling (e.g., 10 CFR Part 55, technical specifications, procedures)	
c. Emergency Plan (e.g., classifications, responsibilities, implementing procedures)	

NRC FORM ES-301N-1			Page 3 of 6
RESEARCH AND TEST REACTOR **OPERATOR LICENSING INDIVIDUAL EXAMINATION REPORT**			
FACILITY NAME:	CANDIDATE DOCKET NUMBER:		
Operating Test			
CATEGORY B: FACILITY WALKTHROUGH			
SYSTEM/TASK TITLE[1]	SYSTEM TYPE[1]	GRADE	Comment Page
1.	MAJ		
2.	EXP		
3.	ESF		
4.	INST		
5.	RDS		
6.	AUX[2]		
7.			
8.			
9.			
10.			

[1]   For Complex facilities a minimum of nine different systems from as many different areas as applicable. For Moderate facilities a minimum of six different systems from as many different areas as applicable. For Simple facilities a minimum of three different systems from as many different areas as applicable.

[2]   For many Moderate and all Simple facilities there are no applicable systems within this area.

NRC FORM ES-301N-1

RESEARCH AND TEST REACTOR
OPERATOR LICENSING INDIVIDUAL EXAMINATION REPORT

FACILITY NAME:	CANDIDATE DOCKET NUMBER:

CATEGORY C: INTEGRATED FACILITY OPERATIONS

INITIAL CONDITIONS:

TURN OVER:

EVENT NUMBER	PERFORM/ DISCUSS	EVOLUTION TYPE	
1			
2			
3			
4			
5			
6			
7			
8			
9			

RESEARCH AND TEST REACTOR
OPERATOR LICENSING INDIVIDUAL EXAMINATION REPORT

FACILITY NAME:	CANDIDATE DOCKET NUMBER:

OPERATING TEST OUTLINE AND GRADING
CATEGORY C: INTEGRATED FACILITY OPERATIONS
COMPETENCY GRADING CHECK SHEET

1. Identification of Cues, Alarms, Annunciators and Trends

		WEIGHT	3.0	2.0	1.0	ACTUAL
a.	Recognize & Acknowledge	0.30	0.90	0.60	0.30	
b.	Locate, Interpret, & Verify Status	0.40	1.20	0.80	0.40	
c.	Prioritize Problems	0.30	0.90	0.60	0.30	

INDIVIDUAL COMPETENCY GRADE

2. Diagnosis of and Response to Conditions

		WEIGHT	3.0	2.0	1.0	ACTUAL
a.	Diagnose Conditions	0.40	1.20	0.80	0.20	
b.	Report Conditions	0.20	0.60	0.40	0.40	
c.	Effect of Actions	0.40	1.20	0.80	0.20	

INDIVIDUAL COMPETENCY GRADE

3. Procedures and Technical Specifications

		WEIGHT	3.0	2.0	1.0	ACTUAL
a.	Recognized Entry Conditions	0.40	1.20	0.80	0.40	
b.	Compliance	0.30	0.90	0.60	0.30	
c.	Reference	0.30	0.90	0.60	0.30	

INDIVIDUAL COMPETENCY GRADE

4. Control Board Basics

		WEIGHT	3.0	2.0	1.0	ACTUAL
a.	Locate and Manipulate Controls	0.33	1.00	0.67	0.33	
b.	Understanding	0.33	1.00	0.67	0.33	
c.	Manual Control	0.33	1.00	0.67	0.33	

INDIVIDUAL COMPETENCY GRADE

RESEARCH AND TEST REACTOR
OPERATOR LICENSING INDIVIDUAL EXAMINATION REPORT

FACILITY NAME:	CANDIDATE DOCKET NUMBER:

OPERATING TEST OUTLINE AND GRADING
COMMENT PAGE

SUBJECT INDEX   COMMENT

ATTACHMENT 1
SYSTEMS FOR OPERATING TESTS

MAJOR SYSTEMS (MAJ)
Primary System
Secondary System
Makeup Water System/Purification System
Pool System(s)
Reactor Design/Construction
Neutron Source

EXPERIMENTAL FACILITIES (EXP)
Beam Tube
In-Core Irradiation
Pneumatic Tube System
In-Pool Irradiation Facilities
Lazy Susan
Thermal Column
Medical Exposure Facilities
Exposure Rooms
Hydraulic Tube System

INSTRUMENTATION SYSTEMS (INST)
Startup Channel (Proportional Counters)
Log-N Channel (Compensated Ion Chambers)
Safety Channel(s) (Uncompensated Ion Chambers)
N^{16} Detectors
Temperature Detectors (RTDs, Thermocouples)
Flow Detectors (Orifices, ΔP Cells)
Control/Safety Rod Drive Systems
Annunciator Systems
Rod Position Indicating Systems
Automatic Rod Positioning Systems

RADIATION DETECTION SYSTEMS (RDS)
Area Radiation Monitoring
Gaseous Radiation Monitoring
Air Particulate Radiation Monitoring
Liquid Effluent Radiation Monitoring
Fission Product Monitoring
Portable Monitors (Ion Chambers and GM Tubes)

ENGINEERED SAFETY FEATURES (ESF)
Reactor Protective System
Core Spray
Pool Makeup/Fill Systems
Containment/Reactor Building Isolation
Anti-Siphoning System

AUXILIARY SYSTEMS (AUX)[1]
Service Air
Reactor Building Air Recirculation
Liquid Waste System
Solid Waste System
Gaseous Waste System
Normal AC Supply
Emergency AC
Emergency DC
Batteries
Motor Generators
Power Inverters
Diesels/Gas Engines

[1]   For many Moderate and all Simple facilities there are no applicable systems within this area.

ATTACHMENT 2
FACILITY LISTING BY TYPE[1]

COMPLEX

1. National Institute of Standards and Technology (NIST), 50-184 (TEST)

2. University of Missouri—Columbia, 50-186 (TANK)

3. Massachusetts Institute of Technology, 50-020 (TANK)

4. Rhode Island Atomic Energy Commission, 50-193 (POOL)

5. University of California, Davis—McClellan Nuclear Research Center, 50-607 (TRIGA)

6. University of Texas, 50-602 (TRIGA)

7. Armed Forces Radiobiology Research Institute, 50-170 (TRIGA)

8. Oregon State University, 50-243 (TRIGA)

9. Texas A&M University, 50-128 (TRIGA)

10. United States Geological Survey (USGS) Department of Interior, 50-274 (TRIGA)

11. Pennsylvania State University, 50-5 (TRIGA)

12. University of Wisconsin, 50-156 (TRIGA)

13. Washington State University, 50-27 (TRIGA)

14. University of Massachusetts—Lowell, 50-223 (POOL)

16. North Carolina State University, 50-297 (PULSTAR)

MODERATE

1. Dow Chemical Company, 50-264(TRIGA)

2. Aerotest Operations, Inc., 50-228 (TRIGA)

3. Kansas State University, 50-188 (TRIGA)

4. Reed College, 50-288 (TRIGA)

5. University of Arizona, 50-113 (TRIGA)

6. University of California at Irvine, 50-326 (TRIGA)

7. University of Maryland, 50-166 (TRIGA)

8. University of Missouri—Rolla, 50-123 (POOL)

9. General Electric Company—Nuclear Test Reactor 50-73 (NTR)

10. Rensselaer Polytechnic Institute, 50-225 (Cr. Ex.)

11. University of Florida, 50-83 (ARGONAUT)

12. University of Utah, 50-407 (TRIGA)

13. Ohio State University, 50-150 (POOL)

14. Worcester Polytechnic Institute, 50-134 (POOL)

15. Purdue University, 50-182 (LOCKHEED)

SIMPLE

1. Idaho State University, 50-284 (AGN-201)

2. Texas A&M University, 50-59 (AGN-201)

3. University of New Mexico, 50-252 (AGN-201)

---

[1] Facilities within this table may be moved between types without requiring a revision to the examiner standards.

EXAMINER STANDARD 302N
ADMINISTRATION OF OPERATING TESTS

A. Purpose

This standard describes the procedure for administering operating tests according to the requirements of 10 CFR 55.45. It assumes that the examiner has prepared for the operating test according to ES-301N.

B. Personnel Present

The number of persons present during an examination should be limited to ensure the integrity of the examination and to reduce distractions to the applicant. According to 10 CFR 55.13(a)(2), a licensed RO or SRO must be present when doing reactivity manipulations. If the number of persons or the noise level in the control room is excessive, the examiner should have the facility staff address the issue.

Except for the licensed RO, no other member of the facility training or operations staff will be allowed to witness an operating test without the permission of the CE. Under **no** circumstances is another applicant allowed to witness an operating test. Operating tests are not to be used as training vehicles for future applicants. Videotaping the administration of initial examinations is not allowed.

Examiners may witness an operating test as part of their training or to audit an examiner administering the operating test. Other observers, such as RTR branch personnel, researchers, or NRC supervisors, may be allowed to observe operating examinations if (1) the CE has approved the request to observe the test and (2) the applicant does not object to the observer's presence.

C. Examination Withdrawals

Occasionally, an applicant will withdraw from the examination just before its start. If this happens, the examiner will request a letter from the facility withdrawing the application of the individual(s). This letter should be addressed to the CE who will forward it to the OLA for inclusion in the applicant's 55 docket file.

In rare instances, an applicant will withdraw after the examination has begun. The examiner will inform the applicant that this is an examination failure, and he or she must reapply following the rules of 10 CFR 55.35.

D. Administration Procedures

1. General

The examiner will use Attachment 1 to brief each candidate before beginning the operating test.

2. Administrative Topics

a. Subcategory A.1

These questions are not intended to duplicate administrative system requirements covered in Categories B and C (e.g., valve lineups, control room data system administration and use). However, the administration of the walkthrough and the scenario often is the best time to ask these questions, in conjunction with the questions associated with Categories B and C.

b. Subcategory A.2

Subcategory A.2 subjects are also best covered during the conduct of tasks or questioning associated with Category B, while walking around the facility in the vicinity of various radiation equipment.

c. Subcategory A.3

The examiner may best evaluate the applicant's emergency plan knowledge by integrating it into a Category C transient discussion or by conducting a Category B task requiring its use. Security may best be evaluated by observing the applicant's demeanor and answers to seemingly innocent questions concerning doors, restrictions on experimenters and operations personnel, and similar topics.

3. Facility Walkthrough

The examiner should encourage the applicant to draw diagrams, flowpaths, and other visual representations. Likewise, the examiner should encourage the applicant to use facility forms, schedules, procedures, and similar materials, as appropriate. In addition, if it clarifies the question, the examiner may present a drawing for interpretation by the applicant.

The examiner should retain any supporting material used during an examination to provide additional documentation in support of a pass or fail determination.

The examiner can improve efficiency of the operating test by integrating discussions required to complete the various categories. This is particularly true for parts of Category C that must be conducted in a discussion format. For example, by postulating an abnormal facility condition, such as a reactor trip/scram, the examiner may include related tasks and associated subject area questions to cover systems in Category B.

The examiner must take sufficient notes during the operating test to document all applicant deficiencies thoroughly. The examiner must cross-reference every comment to a specific task or subject area question.

4. Integrated Facility Operations

Just before starting a scenario, the examiner may review it with the licensed RO. This review should familiarize the licensed RO with the sequence of the scenario events to ensure that it will go as planned with respect to the capabilities and limitations of the facility and the anticipated applicant actions. Precautions should be taken so that the scenarios are not revealed to an applicant before his or her test begins (Form ES-302N-1).

If the applicant acts in a way other than expected, the examiner should note his or her actions (or lack of actions). These notes must provide sufficient information to allow the examiner to judge performance with respect to the K/A competencies described in Form ES-303N-1. Each examiner must determine the best way to document applicant actions. Some examiners record a minute-by-minute account of all key facility events and operator actions that occur, while others only record the candidate's significant actions. It is left to the individual to develop his or her own technique. However, the documentation must provide an adequate basis for making a licensing decision.

The examiner should limit discussions with the applicant during the reactor startup both to maintain realism and to avoid distracting the applicant from operating the facility. The examiner should limit questions asked during the scenario performance to those necessary to assess the candidate's understanding of facility conditions and required operator actions. Even these questions should be deferred until a time (such as during breaks) when the applicant is not operating or closely monitoring the reactor facility. The examiner's followup questions or concerns can often be addressed during a brief question and answer period after each scenario or during the facility walkthrough portion of the operating test if it is done after the operating demonstration.

If the reactor facility becomes inoperable, the CE should discuss the situation with the applicable BC (program manager) so that a decision on the conduct of the operating test can be made. It may be necessary to substitute discussions of transient operating conditions or to reschedule the examinations for a later date.

ATTACHMENTS/FORMS:

Attachment 1, Operating Test Briefing Checklist
Form ES-302N-1, Examination Security Agreement

ATTACHMENT 1
OPERATING TEST BRIEFING CHECKLIST

Part A—Applicable to All Operating Tests

(1)   Senior operator applicants are tested at the level of responsibility of the senior licensed shift position (i.e., shift supervisor, senior shift supervisor, or whatever the position may be titled).

(2)   I am a visitor. Escort responsibilities for ensuring compliance with safety, security, and radiation protection procedures rest with you (the applicant).

(3)   Do not operate facility equipment without appropriate permission.

(4)   Do not hesitate to request clarification of a question during the operating test.

(5)   Frequently I will stop to update my notes to document your performance. The amount of note-taking is not indicative of your level of performance.

(6)   Operating tests are considered open reference. Any reference material in the facility normally available to operators is also available to you. This includes calibration curves, previous log entries, piping and instrument diagrams, calculation sheets, and procedures. However, you are responsible for knowing automatic actions, setpoints and interlocks, operating characteristics, and the immediate actions of emergency and other procedures, as appropriate to the facility.

(7)   There is no specific time limit for the operating test. The exam will take whatever time is necessary to cover the areas selected in the depth and scope required. Typically an RO examination is between 2-1/2 and 3 hours, an SRO-I exam between 3 and 3-1/2 hours, and an SRO-U exam between ½ and 2 hours.

(8)   I am not allowed to reveal the results of the operating test at its conclusion.

(9)   Do not hesitate to request a break during the operating test.

(10)   During the examination you will be evaluated for your actions as if you were the actual watchstander. Please operate the reactor as if you were licensed, with the exception that you should announce your actions, then pause momentarily to give the operator of record time to correct you or stop you, if necessary, before you actually perform the action. In addition, the examiner will be observing that you meet all conditions of your license, (e.g., wearing corrective lenses to perform licensed duties).

Part B—Just before Reactor Startup

Briefing to the candidate and the licensed operator:

(1)   I will not intentionally ask you (the applicant) to perform an action that violates facility regulations or procedures or which places the facility in a hazardous condition. If a requested action meets one of these conditions, then you (the applicant), or you (the operator) should immediately inform me. If my intent was to find out whether you (the applicant) would perform such an act, I will phrase the question in some manner other than requesting that the act be performed.

(2)   My presence does not alter the normal chain of command during the examination. You (the applicant) should make all reports and obtain all permissions that would normally be required. All directions to the applicant will come from the responsible supervisor following the facility administrative procedures. The examiner will only question and make requests of the supervisor.

(3)   I have not altered the setpoints or calibrations of any instrument nor have I manipulated any controls.

(4)   It is your (licensed operator) responsibility to step in and take control of the reactor any time there is an unsafe condition or if you think that the reactor will shut down if conditions are not corrected. However, you may not provide any coaching or cuing to the applicant.

FORM ES-302N-1
EXAMINATION SECURITY AGREEMENT

Preexamination Security Agreement

I _____ agree that I will not knowingly divulge any information
       [Print Name]
concerning the operating tests scheduled for _____ to any unauthorized
                                                [Print Date]
persons. I understand that I am not to participate in any instruction involving the reactor
operator or senior reactor operator applicants scheduled to be administered the above
operating test from now until after the examination has been administered.

_____
Signature/Date

Postexamination Security Agreement

I _____ did not, to the best of my knowledge, divulge any
       [Print Name]
information concerning the operating tests administered on _____ to any
                                            [Print Date]
unauthorized persons. I did not participate in providing any instruction to the reactor operator
and senior reactor operator applicants who were administered the operating test from the time I
was allowed access to the operating test.

_____
Signature/Date

EXAMINER STANDARD 303N
GRADING OPERATING TESTS

A. Purpose

This standard describes the procedures for grading operating tests given according to 10 CFR 55.45. It includes methods for documenting all aspects of the operating test, making pass/fail recommendations, and reviewing the documentation to ensure a quality product.

B. General Evaluation Guidelines

S   Satisfactory Working Knowledge and Understanding

The applicant may have had some minor difficulty relating to system interactions. His or her operation of equipment was very good, although he or she may have shown some hesitation while discussing or performing some tasks. Overall, the applicant seemed familiar with the equipment and procedures.

U   Unsatisfactory Working Knowledge and Understanding

The applicant had difficulty answering questions in depth or in relating the interactions of systems. The applicant showed a lack of familiarity with the equipment or the procedures. The applicant's answers were inaccurate or incomplete. The applicant displayed unfamiliarity with the subject or system, as shown by hesitant answers, inability to find information, inability to find control board indications or controls, or the lack of knowledge of procedural steps to operate systems.

Detailed notes must support all deficiencies stating the particular action or response displaying the deficiency. Sufficient knowledge deficiencies in a common area may result in an unsatisfactory evaluation for that area.

When documenting deficiencies, general statements such as "did not know primary coolant system" are inadequate. The examiner must note specific items and the significance of each deficiency to reactor safety or the health and safety of the public or reactor staff.

The examiner must document all Unsatisfactory grades and all numerical grades of 1 or 2 for failed competencies by providing the following (to the extent applicable):

- the question asked by the examiner
- the incorrect answer from the applicant
- the knowledge or ability the applicant lacks
- the consequences of the incorrect answer
- the correct answer

C. Specific Instructions for Completing Form ES-303N-1

The examiner must use Form ES-301N-1 (and all applicable attachments) to grade the test according to the following instructions.

1. Category A, Administrative Topics (Form ES-301-1, page 2)

The examiner must evaluate the applicant's performance for each administrative topic discussed or performed. All comments should be entered on Form ES-301N-1, page 6 (or higher). The examiner will review all comments concerning a topic, make a SAT/UNSAT determination for the topic, and document the grade by placing an **S** or **U** in the appropriate block on ES-301N-1, page 2. This process is then repeated for each topic covered in the category. For topics not covered during the test, the examiner should indicate that no grade is to be assigned by placing a dash or the letters N/E (Not Examined) in the block; this will clearly show that the block was not missed.

After grading all applicable topics in the category, the examiner will determine the overall grade for Category A by dividing the number of satisfactory topics by the total number of topics examined. The applicant must be graded as satisfactory in 75 percent of the topics to receive an overall satisfactory grade. This grade is then placed in the *Category A* block in the *Operating Test Summary* section of the ES-301N-1 cover page.

2. Category B, Facility Walkthrough (Form ES-301-1, page 3)

The examiner must indicate the systems selected and the tasks conducted in the appropriate columns of Form ES-301-1, page 3. The examiner must then evaluate the applicant's performance for each of those systems.

All comments should be entered on Form ES-301N-1, page 6. The examiner will review all comments on the system, make a SAT/UNSAT determination, and document the grade by placing an **S** or **U** in the appropriate block on ES-301N-1, page 3. This process is then repeated for all systems examined.

Finally, the examiner will determine the overall category grade by calculating the ratio of systems passed to systems evaluated. To receive a grade of "Satisfactory," the applicant must receive a satisfactory evaluation in at least 75 percent of the systems examined.

For example, if a candidate on a reactor facility examination received one unsatisfactory system grade and seven satisfactory grades, he or she would receive an **S** for the category based on a grade of 7/8 (or 87 percent).

3. Category C, Scenario Events (Form ES-301N-1, pages 4 and 5)

Before grading the operating test, the examiner must enter any scenario revisions made during the test onto page 6 of the applicable ES-301N-1 so that the evaluation form accurately reflects the events actually tested. The final product must be a clear, legible reproduction of the actual events that occurred at the facility during the operating test. The NRC will expunge any rough notes or nonpertinent comments from the final version sent to the candidate.

The examiner will grade candidate performance by using the worksheet (Form ES-303N-1) and then transcribing the grade onto page 4 of Form ES-301N-1.

The worksheet contains several competencies for measuring operator performance. Each competence has several rating factors, with scales of 1 through 3, further defining the competence. The rating scales have definitions, or "behavioral anchors," at each point along the scale. These anchors aid the examiner in assigning a value based on the applicant's actions. Each rating factor is also assigned a specific weight representing the relative importance of that factor. The weight factor, in conjunction with the value assigned to the rating factor by the examiner, yields a numerical measure of the applicant's performance on that rating factor. Combining all the rating factor numerical evaluations for all competencies will yield the applicant's overall grade for the integrated facility operation portion of the operating test.

The examiner must complete the following to evaluate the applicant's performance on the integrated facility operation portion of the operating test:

a.   Review and Categorize Notes. Review the notes taken while administering the examination. Label the documented actions and behaviors with the number and letter of the rating factor they most accurately reflect. Each comment that reflects negatively on the candidate's performance should be coded, whether or not it results in a grade of 1 or 2 for that rating factor.

b.   Evaluate the Candidate on the Rating Factors. After categorizing the actions and behaviors, evaluate the applicant's performance by completing page 4 of Form ES-301N-1. Form ES-303N-1 gives the examiner guidance for evaluating each rating factor. For each rating factor, circle the integral rating value corresponding to the rating value (1–3) that most accurately reflects the applicant's performance. (As discussed in ES-301N, *Control Board Operations* is optional for SRO-U applicants.)

c.   Document Low Ratings. All ratings of 1 must be justified by documenting the specific actions or behaviors that warranted the evaluation and their consequence. Additionally, ratings of 2 must be documented for any competency that contributes to a candidate failing Category C. Provide this documentation directly on page 6 of Form ES-301N-1.

d.   Compute Competence Grades. For each competence, sum the rating grades on ES-301N-1, page 4, to compute a final competence grade. (Final competence grades will range between 1 and 3.) Record the final competence grades in the *Total* blocks of the form.

e.   Compute Overall Competence Rating. If the applicant receives a score greater than 1.8 in all competencies examined, then the examiner will enter an overall grade of Satisfactory (SAT or **S**) in the Category C block on the cover page of Form ES-301N-1. (NOTE: For an SRO-U applicant, if Competence 5, *Control Board Operations*, is evaluated, it should be factored into the applicant's final grade.)

4. Operating Test Comments (Form ES-301N-1, page 6)

The examiner must document all weaknesses found during the operating test on page 6 of Form ES-301N-1. The examiner will enter the specific subject (e.g., A.3.c, B.1.3.C) on the left side of the form and the specific comment to the right. The *Comment Page No.* blocks are used to cross-reference comments between the form and the attachments. Paragraph D.5 contains additional instructions regarding the documentation of Category C comments.

Any figures, drawings, flowcharts, forms, or the like that the applicant used during the operating test may be used in documenting his or her performance. They should be appropriately marked and cross-referenced to applicable comments on Form ES-301N-1 and attached to the examination package for retention.

5. Cover/Summary Page

The cover page of Form ES-301N-1 summarizes all of the pertinent information about the applicant, the facility, and the examination.

The examiner will fill in the applicant's name, docket number, the facility for which the license is sought, the type of examination, and the facility description.

Next the examiner will enter the name of the examiner who conducted the operating test and the date on which it was given. The grading summary blocks must be completed according to the instructions in this standard by the examiner who administered the operating test.

The examiner who conducted the operating test will enter his or her pass, fail, or waive recommendation in the appropriate block. The examiner may make a pass recommendation only if **all** summary blocks of the operating test contain SAT (**S**) grades or N/E. The examiner will then sign as the test administrator.

Once completed, the entire examination package is given to the CE.

ATTACHMENTS/FORMS:

Form ES-303N-1, RO/SRO Competency Grading Guidance for Integrated Facility Operations

Form ES-303N-1 RO/SRO Competency Grading Guidance for Integrated Facility Operations

1. IDENTIFICATION OF CUES, ALARMS, ANNUNCIATORS, AND TRENDS

This competency involves the ability to *notice and acknowledge* cues, alarms, annunciators, and trends. It includes the ability to prioritize one's attention in keeping with the severity/importance of facility status information and the ability to *interpret correctly and verify* that signals are *consistent with facility/system conditions* (with the use of Alarm Response Procedures, as appropriate). The competency deals strictly with the understanding and interpretation of annunciators and alarm signals and therefore does not include knowledge of, or the ability to diagnose, overall facility/system status based on other indications of facility/system status or condition(s).

DID THE CANDIDATE:

(A) RECOGNIZE and ACKNOWLEDGE cues, alarms, annunciators, and trends?

3	2	1
Consistently and timely recognition/ acknowledgment	Some lapses in awareness	Failed to recognize/acknowledge cues, alarms, annunciators, or trends

(B) Correctly INTERPRET and VERIFY that information is consistent with facility/system conditions (including the use, when necessary, of the Alarm Response Procedures)?

3	2	1
Consistently and effectively	Some inaccuracies in interpretation/ verification of signals	Inaccuracies resulting in facility degradation; poor use of ARPs

(C) ATTEND to ANNUNCIATOR/ALARM SIGNALS, CUES, and TRENDS in order of importance/severity?

3	2	1
Yes in all cases	Some inaccuracies/oversights	Did not prioritize attention to facility status information; inattentive to important alarms

2. DIAGNOSIS OF AND RESPONSE TO CONDITIONS

This competency involves the ability to *diagnose* facility conditions to recognize and *report* these to mitigate out-of-spec conditions. Included is the *knowledge* of the impact of continued operation under out-of-spec conditions and the *knowledge* of the effects of mitigating actions. This competency also includes correctly reporting these conditions to the appropriate individuals, departments, or agencies.

DID THE CANDIDATE:

(A) Correctly DIAGNOSE facility conditions based on control room indications?

3	2	1
Diagnoses were accurate	Some errors/difficulties in diagnoses	Faulty diagnoses adversely impacted facility status

(B) Correctly REPORT conditions out of specification, diagnoses of conditions, and continuing status?

3	2	1
Accurate and timely reports provided	Some lapses in making reports	Delayed or faulty reports adversely impacted mitigation or emergency response

(C) Demonstrate an understanding of the EFFECTS of continued operation and responsive actions and how they AFFECTED FACILITY/SYSTEM CONDITIONS?

3	2	1
Understood the effects of continued operation and responsive actions	Some misunderstandings with no significant impact	Inadequate knowledge of effects resulted in facility degradation

Form ES-303N-1 RO/SRO Competency Grading Guidance
for Integrated Facility Operations

3.   COMPLIANCE/USE OF PROCEDURES AND TECHNICAL SPECIFICATIONS

This competency involves the ability to refer to and comply with normal, abnormal, emergency and administrative procedures in a timely manner (i.e. in sufficient time to avoid adverse impacts on facility status). Included is the ability to recognize procedure entry conditions, and carry out required actions correctly, and to recognize and comply with required LCO/Action Statements.

DID THE CANDIDATE:

(A)   RECOGNIZE entry condition requirements for Normal, Abnormal, Emergency Procedures, and Technical Specifications?

3	2	1
Consistently and timely recognition / acknowledgment of conditions requiring procedure driven actions	Some difficulties/oversights made in understanding procedure entry conditions	Significant errors in important instances resulted in missed requirements or facility degradation

(B)   COMPLY WITH procedures (including precautions and limitations) in an accurate, timely manner?

3	2	1
Accurate and timely compliance	Few errors; corrections made in sufficient time to avoid adverse impact	Many significant errors; required excessive assistance

(C)   REFER to the appropriate procedure/Technical Specification in a timely manner?

3	2	1
Quickly located appropriate procedures/Tech specs.	Some difficulties/oversights made in referring to appropriate procedures	Problems/failures in referring to appropriate procedure(s) in important instances

4.   CONTROL BOARD BASICS

This competency involves the ability to locate and manipulate controls to attain a desired facility/system response/condition. Included is the ability to take manual control of automatic functions, when appropriate.

DID THE CANDIDATE:

(A)   LOCATE/MANIPULATE CONTROLS effectively and accurately?

3	2	1
Went directly to appropriate controls and manipulated accurately in a timely manner	Some hesitancy/difficulty in locating or manipulating controls but efficiently mitigated any resulting consequences	Demonstrated inability to locate controls without assistance, or major system perturbation(s) resulting from improper manipulations.

(B)   Demonstrate an understanding of how his or her ACTIONS (or inaction) AFFECT PLANT and SYSTEM CONDITIONS?

3	2	1
Understood the effect of actions on plant and systems	Some misunderstanding of effect of actions on plant and systems	Appeared to act without knowledge of or regard for effect on plant and systems

(C)   Take MANUAL CONTROL of automatic functions when appropriate?

3	2	1
Took manual control as appropriate	Some delays; some as appropriate before overriding automatic functions	Depended on automatic actions; required prompting to take manual control

EXAMINER STANDARD 401N
PREPARATION OF WRITTEN EXAMINATIONS

A. Purpose

This standard specifies the requirements, procedures, and guidelines for the preparation of written examinations for the licensing of RO and SRO applicants at research and test reactor facilities.

B. Background

The content of the written licensing examinations is contained in 10 CFR 55.41, "Written Examinations: Operators," and 10 CFR 55.43, "Written Examinations: Senior Operators." Each examination will include a representative selection of questions on the K/As needed to perform licensed duties. The written examination is administered in three sections, as described in Section D of this standard.

C. Examination Preparation

Examiners will prepare written examinations according to the guidelines and instructions summarized below.

1. General Guidelines

a. Prepare the examination so that an applicant capable of safely operating the facility can complete and review the examination within the time allotted, achieving a grade of at least 70 percent in each category. When creating the examination, estimate 2 minutes per response.

b. Avoid excessive duplication of questions. Do not repeat more than 25 percent of the questions from the last NRC-prepared examination administered at the facility.

c. All questions will be in the multiple choice (preferred) or matching format. Matching questions should be limited to a total value of 2 points per question. All multiple choice questions will have four choices.

d. It is desirable to develop questions that hypothesize events or circumstances leading to events, thereby examining the applicants' analytical abilities and knowledge of corrective actions.

e. Questions should contain a representative sample from among the following 21 items, to the extent applicable to the facility and the exam level (RO):

    (1)   fundamentals of reactor theory, including fission process, neutron multiplication, source effects, control rod effects, criticality indications, reactivity coefficients, and poison effects.

    (2)   general design features of the core, including core structure, fuel elements, control rods, core instrumentation, and coolant flow

    (3)   mechanical components and design features of the reactor primary system.

(4)   secondary coolant and auxiliary systems that affect the facility

(5)   facility operating characteristics during steady-state and transient conditions, including coolant chemistry, causes and effects of temperature, pressure and reactivity changes, effects of load changes, and operating limitations and reasons for these operating characteristics

(6)   design, components, and functions of reactivity control mechanisms and instrumentation

(7)   design, components, and functions of control and safety systems, including instrumentation, signals, interlocks, failure modes, and automatic and manual features

(8)   components, capacity, and functions of emergency systems

(9)   shielding, isolation, and containment design features, including access limitations

(10)   administrative, normal, abnormal, and emergency operating procedures for the facility

(11)   purpose and operation of radiation monitoring systems, including alarms and survey equipment

(12)   radiological safety principles and procedures

(13)   procedures and equipment available for handling and disposal of radioactive materials and effluents

(14)   principles of heat transfer thermodynamics and fluid mechanics

(15)   conditions and limitations in the facility license

(16)   facility operating limitations in the technical specifications and their bases

(17)   facility licensee procedures required to obtain authority for design and operating changes in the facility

(18)   radiation hazards that may arise during normal and abnormal situations, including maintenance activities and various contamination conditions

(19)   assessment of facility conditions and selection of appropriate procedures during normal, abnormal, and emergency situations

(20)   procedures and limitations involved in initial core loading, alterations in core configuration, control rod programming, and determination of various internal and external effects on core reactivity

(21)   fuel-handling facilities and procedures

2. Use of Reference Material

The examiner is expected to use the following reference materials when preparing the site-specific, written operator licensing examination:

a.   Facility reference material provided per ES-201N, especially facility learning objectives used in the applicants' training program, if available.

b.   NRC questions previously used at the facility or at similar facilities, as applicable.

c.   Licensee event reports, safety evaluation reports, information notices, current industry and facility problems, and facility question banks. Questions drawn from the facility's bank will not exceed 10 percent of the questions in that bank.

d.   The examiner may use reference material (i.e., diagrams, sketches, portions of facility procedures) as attachments to the written examination and ask applicants to identify components and other items on the attachments. The examiner will ensure that any reference material used in the examinations is easy to read, clearly marked, and provides an effective and objective way for the applicant to display knowledge.

3. Examination Assembly

The examiner will produce several copies of the examination, along with a cover sheet stating whether the examination has an answer sheet (master copy and facility review copies) or not (applicant copies). In addition, the examiner will produce several copies of an examination handout similar to Attachment ES-401N-1 to this standard. This handout will contain a cover sheet for the applicant to sign, a copy of the "Policies and Guidelines for Taking NRC Written Examinations," an equation sheet, answer sheets, and any tables, drawings, or graph paper necessary to support the examination. The applicant must sign the cover sheet and submit the handout before leaving the examination area.

D. Examination Structure

There are minimal differences between the KSAs required of RO and SRO applicants that can be tested in a written examination format. Therefore, the NRC only writes only one examination for both RO and SRO-I applicants. Because SRO-U candidates have already passed an RO-level written examination, they are not required to take another written examination. Differences in the KSAs required of an SRO versus an RO are better evaluated on the operating exam (per ES-301N). Modifying an SRO-I exam to include questions that relate directly to SRO duties and responsibilities is permissible, but the operating exam serves as a more suitable forum for evaluating such topics.

Because research and test reactor facilities vary widely in complexity, the structure of the written examination will be based on the three classifications identified in ES-301N, Complex, Moderate, or Simple. Attachment 2 to ES-301N is a list of research and test reactor facilities by classification.

The examination will contain the following three categories:

A   Reactor Theory, Thermodynamics, and Facility Operating Characteristics

B   Normal and Emergency Operating Procedures and Radiological Controls

C   Facility and Radiation Monitoring Systems

The minimum number of questions contained in each of these categories depends upon the facility type.

Complex reactor facility examinations should have a minimum of 20 questions per category. Moderate reactor facility examinations should have a minimum of 15 questions per category, and Simple reactor facility examinations should have a minimum of 10 questions per category.

The written examination will be divided into the three categories identified and described below.

Category A—This category contains questions relating to basic nuclear theory, reactor behavior, and processes that take place in a reactor. Thermodynamics should be limited only to those areas that are applicable to the facility's operations and should be related to pertinent heat transfer and fluid flow processes that are observable at the facility. Questions should also relate to the manner in which power, reactivity, rod worth, and other parameters of the facility change in response to rod manipulation, core burnup, heatup, experiment insertion, or other activities associated with the operation of the facility. Questions on facility behavior as seen on recorder traces because of these activities should be included.

Category B—This category contains questions on the procedures for operation of the reactor and auxiliary systems, including administrative controls pertinent to the RO position. Candidates must display complete understanding of the immediate action steps and the bases associated with abnormal and emergency operating procedures. Familiarity and understanding of technical specifications and operating, surveillance, and maintenance procedures to the extent applicable to the RO position should also be evaluated. This category will also include questions on radiation hazards, radiological safety practices, and facility and Federal regulations (e.g., 10 CFR Part 20) for the identification and control of radiation hazards and radioactive material handling and releases.

Category C—This category contains questions on the design, construction, operation, and interrelationships of the systems most directly associated with reactor safety, such as rod control, emergency power, or core flooding. Questions on the characteristics, operation, and interrelationships of nuclear and process instrumentation and control systems are also to be included, with an emphasis on safety-related devices. Investigating candidate knowledge of auxiliary systems and instrumentation is allowable, but the bulk of the questions will be associated with areas important to safety. This category shall also include questions associated with radiation monitoring systems and detectors, including both fixed and portable equipment. These questions will evaluate the candidate's knowledge of the operating characteristics, limitations, and applications of the equipment as appropriate to the facility.

The goal for the relative weight of each category in the examination should be approximately 33 percent of the total examination worth. However, since the relative importance of safety and emergency systems varies significantly from one size and type of research and test reactor to another and 10 CFR 55.41 allows flexibility in the selection of questions "to the extent applicable to the facility...," examiners experienced in the operation and examination activities associated with research and test reactors can modify the weighting of the examination categories based on their professional judgement. In any case, the relative weight of any one category should be no less than 20 percent and no more than 40 percent of the total examination worth.

E. NRC Quality Assurance Reviews

After preparing the examination, the author should independently review each examination question for content and wording. When reviewing, the author should place himself or herself in the position of the applicant by attempting to answer the questions without using reference material or the answer key. In addition, he or she should ensure that the conditions and requirements posed in the question are complete and unambiguous and that all necessary information is provided, all unnecessary information is deleted, and the intended answer clearly follows from what is asked in the question.

If time and resources permit, another NRC examiner may independently review the examination using the guidance above. A second reviewer is especially important for examiners in training, newly certified examiners, and examination authors who are not the CE.

All reviewers should review the examination using Form ES-401N-1, which is included at the end of this standard.

If the examination review identifies major changes in structure or content (e.g., distractor or question replacements), the CE is responsible for consolidating the comments from all reviewers and submitting one set of comments to the author. If the examination conforms with the requirements of the "Examiners' Handbook" and this standard and was transmitted electronically to the staff, the CE may make the desired changes directly without involving the author of the examination.

After the necessary changes have been made, the CE will review the final version of the examination for completeness and clarity.

ATTACHMENTS/FORMS:

Attachment ES-401N-1, Sample Written Examination Cover Sheet, NRC Rules for Written Examination Administration, Equation Sheet, and Answer Sheets
Form ES-401N-1, Written Examination Quality Assurance Checkoff Sheet

ATTACHMENT ES-401N-1
WRITTEN EXAMINATION COVER SHEET

U.S. NUCLEAR REGULATORY COMMISSION
RESEARCH AND TEST REACTOR LICENSE EXAMINATION

FACILITY: _____

REACTOR TYPE: _____

DATE ADMINISTERED: _____

CANDIDATE: _____

INSTRUCTIONS TO CANDIDATE:

Answers are to be written on the answer sheets provided. Attach all answer sheets to the examination. Point values are indicated in parentheses for each question. A 70 percent in each category is required to pass the examination. Examinations will be collected 3 hours after the examination starts.

CATEGORY VALUE	% OF TOTAL	CANDIDATE'S SCORE	% OF CATEGORY VALUE	CATEGORY
20.00	33.3	_____	_____	A. REACTOR THEORY, THERMODYNAMICS, AND FACILITY OPERATING CHARACTERISTICS
20.00	33.3	_____	_____	B. NORMAL AND EMERGENCY OPERATING PROCEDURES AND RADIOLOGICAL CONTROLS
20.00	33.3	_____	_____	C. FACILITY AND RADIATION MONITORING SYSTEMS
60.00		_____	_____ %	TOTALS
		FINAL GRADE		

All work done on this examination is my own. I have neither given nor received aid.

_____

Candidate's Signature

ATTACHMENT ES-401N-1 (Continued)
NRC RULES FOR WRITTEN EXAMINATION ADMINISTRATION

(1)   Cheating on the examination will result in an automatic denial of your application and could result in more severe penalties.

(2)   After the examination has been completed, you must sign the statement on the cover sheet indicating that the work is your own and that you have not received or given assistance in completing the examination. This must be done after you complete the examination.

(3)   Restroom trips are to be limited and only one applicant at a time may leave. You must avoid all contact with anyone outside the examination room to avoid even the appearance or possibility of cheating.

(4)   Print your name in the blank provided in the upper right corner of the examination cover sheet and each answer sheet.

(5)   Mark your answers on the answer sheet provided. **USE ONLY THE PAPER PROVIDED AND DO NOT WRITE ON THE BACK SIDE OF THE PAGE.**

(6)   The point value for each question is indicated in brackets after the question.

(7)   If the intent of a question is unclear, ask questions of the proctor only.

(8)   When turning in your examination, assemble the cover sheet and answer sheets.

(9)   To pass the examination, you must achieve a grade of 70 percent or greater in each category.

(10) There is a time limit of 3 hours for completion of the examination.

(11) When you have completed and turned in your examination, leave the examination area (DEFINE THE AREA). If you are observed in this area while the examination is still in progress, your license may be denied or revoked.

ATTACHMENT ES-401N-1 (Continued)

EQUATION SHEET

$$\dot{Q} = \dot{m}c_p \, \Delta T = \dot{m} \, \Delta H = UA \, \Delta T \qquad P_{max} = \frac{(\rho-\beta)^2}{2\,\alpha(k)\ell} \qquad \ell^* = 1 \times 10^{-4} \text{ second}$$

$$\lambda_{eff} = 0.1 \text{ seconds}^{-1} \qquad SCR = \frac{S}{-\rho} \approx \frac{S}{1-K_{eff}} \qquad \begin{array}{c} CR_1(1-K_{eff_1}) = CR_2(1-K_{eff_2}) \\ CR_1(-\rho_1) = CR_2(-\rho_2) \end{array}$$

$$SUR = 26.06\left[\frac{\lambda_{eff}\rho}{\beta-\rho}\right] \qquad M = \frac{1-K_{eff_0}}{1-K_{eff_1}} \qquad M = \frac{1}{1-K_{eff}} = \frac{CR_1}{CR_2}$$

$$P = P_0\,10^{SUR(t)} \qquad P = P_0\,e^{\frac{t}{T}} \qquad P = \frac{\beta(1-\rho)}{\beta-\rho}\,P_0$$

$$SDM = \frac{(1-K_{eff})}{K_{eff}} \qquad T = \frac{\ell^*}{\rho-\bar{\beta}} \qquad T = \frac{\ell^*}{\rho} + \left[\frac{\bar{\beta}-\rho}{\lambda_{eff}\rho}\right]$$

$$\Delta\rho = \frac{K_{eff_2} - K_{eff_1}}{k_{eff_1} \times K_{eff_2}} \qquad T_{\frac{1}{2}} = \frac{0.693}{\lambda} \qquad \rho = \frac{(K_{eff}-1)}{K_{eff}}$$

$$DR = DR_0\,e^{-\lambda t} \qquad DR = \frac{6\,CiE(n)}{R^2} \qquad DR_1 d_1^2 = DR_2 d_2^2$$

DR—Rem; Ci—curies; E—Mev; R—feet

$$\frac{(\rho_2-\beta)^2}{Peak_2} = \frac{(\rho_1-\beta)^2}{Peak_1}$$

---

1 Curie = 3.7×10^{10} dis/sec   1 kg = 2.21 lbm

1 Horsepower = 2.54×10^3 BTU/hr   1 Mw = 3.41×10^6 BTU/hr

1 BTU = 778 ft-lbf   °F = 9/5 °C + 32

1 gal (H$_2$O) ≈ 8 lbm   °C = 5/9 (°F - 32)

c_p = 1.0 BTU/hr/lbm/°F c_p = 1 cal/sec/gm/°C

ATTACHMENT ES-401N-1 (Continued)

Section A R Theory, Thermo, and Facility Characteristics   Page 4

A.1a α β γ n ___   A.7   a b c d ___

A.1b α β γ n ___   A.8   a b c d ___

A.1c α β γ n ___   A.9   a b c d ___

A.1d α β γ n ___   A.10 a b c d ___

A.2   a b c d ___   A.11 a b c d ___

A.3a 1 2 3 4 5 6 7 ___   A.12 a b c d ___

A.3b 1 2 3 4 5 6 7 ___   A.13 a b c d ___

A.3c 1 2 3 4 5 6 7 ___   A.14 a b c d ___

A.3d 1 2 3 4 5 6 7 ___   A.15 a b c d ___

A.4   a b c d ___   A.16 a b c d ___

A.5   a b c d ___   A.17 a b c d ___

A.6   a b c d ___   A.18 a b c d ___

ATTACHMENT ES-401N-1 (Continued)

<u>Section B Normal/Emerg. Procedures & Rad Con</u>   Page 5

B.1   a b c d ___   B.9c NR Double NA ___

B.2a 1 2 3 4 5 6 7 8 9 ___   B.9d NR Double NA ___

B.2b 1 2 3 4 5 6 7 8 9 ___   B.10a Water Air Structural Fission ___

B.2c 1 2 3 4 5 6 7 8 9 ___   B.10b Water Air Structural Fission ___

B.2d 1 2 3 4 5 6 7 8 9 ___   B.10c Water Air Structural Fission ___

B.2e 1 2 3 4 5 6 7 8 9 ___   B.10d Water Air Structural Fission ___

B.3   a b c d ___   B.11 a b c d ___

B.4   a b c d ___   B.12 a b c d ___

B.5   a b c d ___   B.13 a b c d ___

B.6   a b c d ___   B.14a ___ days

B.7   a b c d ___   B.14b ___ weeks

B.8a SL LSSS LCO ___   B.14c ___ months

B.8b SL LSSS LCO ___   B.14d ___ months

B.8c SL LSSS LCO ___   B.15 a b c d ___

B.8d SL LSSS LCO ___   B.16 a b c d ___

B.9a NR Double NA ___   B.17 a b c d ___

B.9b NR Double NA ___

ATTACHMENT ES-401N-1 (Continued)

Section C Facility and Radiation Monitoring Systems　　　　　Page 6

C.1 a b c d ___

C.2 a b c d ___

C.3 a b c d ___

C.4 a b c d ___

C.5 a b c d ___

C.6 a b c d ___

C.7 a b c d ___

C.8 a b c d ___

C.9a 1 2 3 4 ___

C.9b 1 2 3 4 ___

C.9c 1 2 3 4 ___

C.9d 1 2 3 4 ___

C.10a Open Closed Throttled ___

C.10b Open Closed Throttled ___

C.10c Open Closed Throttled ___

C.10d Open Closed Throttled ___

C.10e Open Closed Throttled ___

C.10f Open Closed Throttled ___

C.11a Open Shut On Off ___

C.11b Open Shut On Off ___

C.11c Open Shut On Off ___

C.11d Open Shut On Off ___

C.11e Open Shut On Off ___

C.11f Open Shut On Off ___

C.11g Open Shut On Off ___

C.11h Open Shut On Off ___

C.12 a b c d ___

C.13 a b c d ___

C.14 a b c d ___

C.15 a b c d ___

C.16 a b c d ___

C.17 a b c d ___

NRC FORM ES-401N-1

WRITTEN EXAMINATION QUALITY ASSURANCE CHECKOFF SHEET

Examination Package No. 50-[xxx]/OL-YY-[xx], Facility Name, Month Year

Author:	Date of Examination:	
[Name]	[Date]	
	Author (Check)	Peer/CE/BC Review (Check)
Objectivity and clarity of questions. Answers clear and concise on answer key.		
Questions and answers technically accurate and applicable to facility.		
Proper level of knowledge (RO).		
No double jeopardy questions.		
10 CFR 55.41 (RO) sampling is appropriate.		
No question worth more than 2 points.		
Total point value correct for facility classification (I, II, or III)		
ES-401.C.1.(a–d) Examination question format criteria met. (Less than 25 percent overlap of questions from last examination. No true/false questions. All multiple choices questions have four choices with one correct answer. Matching & Completion questions have only one correct answer and are limited to 2 points.		
Author Review:	Date: / /07	
Peer/CE/BC Review (Not Required)	Date: / /07	

EXAMINER STANDARD 402N
ADMINISTRATION OF WRITTEN EXAMINATIONS

A. Purpose

This standard specifies the requirements and procedures for administering the written examination at a research and test reactor facility.

B. Examination Facilities

(1)   The licensee is responsible for providing a room that is suitable for administering the written examination. The examination room should have nearby restroom facilities to enable the examiner and/or proctor to maintain security of the written examination. The NRC's policies regarding the written examination facility and other preparations for administering the written examination are listed in Enclosure 2 in the corporate notification letter (refer to Attachment 1 to ES-201N).

(2)   The CE will evaluate the examination room for adequacy. The applicants should not have access to any reference material that is not approved by the CE. The CE will not begin the examination until he or she is satisfied with the room where the exam is going to be given.

C. Proctoring the Examinations

(1)   The CE will ensure that the examination is proctored at all times. For large numbers of applicants, the CE may determine that he or she needs assistance proctoring the written examination. If proctors are needed, the CE should request assistance from other examiners or other responsible NRC employees.

The CE will brief all proctors on their responsibilities before the examinations are distributed. Proctors shall not engage in any activities that may divert their attention from the applicants and possibly cause the examination to be compromised.

(2)   When an applicant asks for clarification of a question on the written examination, the proctor should write down the applicant's question, and in a case where the proctor is not a certified examiner, present the question to a certified examiner for resolution. When an examiner is not sure how to respond to a question without giving away the answer, the examiner will refer the question to the CE. If there is no way to clarify a question without giving away the answer, the proctor will so inform the applicant.

The proctor(s) will document all questions regarding specific written examination test items on the master copy of the examination. The CE will use the comments for future reference in resolving facility comments and grading conflicts.

When responding to questions, the proctor should be alert for indications that an applicant is unfamiliar with the terminology used in the examination. The proctor will ask the CE to determine the correct terminology and announce it to all the applicants taking the examination.

All question changes or clarifications will be called to the attention of all the applicants. Changes made to questions during the examination should be made on the master copy and on the copy provided to the facility staff.

D. Examination Administration Procedure

The CE will insure that the written examination is administered as follows:

(1)   Verify each applicant's identity against the examination assignment sheet (see Form ES-201N-1). Any errors or absences will be resolved with the facility staff, and the assignment sheet will be updated as required.

The CE will request the facility licensee to withdraw the application of any individual not taking the examination. This withdrawal will be formalized by sending a letter to the BC/PD.

Inspect the examination room and either approve, have removed or have covered any reference material.

(2)   A proctor (whether the CE, a certified examiner, or a non-examiner) will do the following:

a.   Inform the applicants that they may use calculators to complete the examination. Define the examination room for the applicants.

b.   Pass out the examinations, answer sheets, and handouts and instruct the applicants not to review the examination until told to do so.

c.   Brief the applicants on the rules and guidelines that will be in effect during the written examination by directing them to follow along with the instructions directly beneath the examination cover sheet. Read the first two policies **verbatim**.

d.   Ask the applicants to verify the completeness of their examination by checking each page.

e.   After answering any questions that the applicants may have regarding examination policies, start the examination and record the time.

f.   Periodically advise the applicants of the examination time remaining.

g.   Ensure that each applicant signs the cover sheet when turning in his or her examination and ensure that all answer sheets are included.

h.   Remind the applicants to leave the examination area.

(3)   When all of the written examinations are completed, the CE may conduct an examination review with the facility licensee staff as described in Section E below.

E. Facility Staff Review of the Written Examination

After the last applicant completes the examination, the CE will meet with all proctors and ensure that the master copy of the examination is updated with all of the changes made to the questions and/or answers while the examinations were being administered.

The CE will provide a copy of the master examination to the facility licensee staff and answer any questions the staff may have regarding the NRC's examination review and comment process. ES-201N, Enclosure 4, provides detailed guidelines and instructions for this review process.

ATTACHMENTS/FORMS:

NONE

EXAMINER STANDARD 403N
GRADING WRITTEN EXAMINATIONS

A. Purpose

This standard describes the requirements and procedures for resolving the facility licensee's comments on the written examination, grading the examination, and conducting a quality assurance review of the graded written examination.

B. Resolving Facility Comments

(1)   The examiner grading the examination will analyze each facility comment and document the reason that the question was changed or the reason that the comment was not accepted.

(2)   If there are two correct answers to a multiple choice question, both answers will be accepted as correct. However, if three or more answers could be considered correct, the question will be deleted from the examination.

(3)   If 10 percent or more of the questions are deleted as a result of facility licensee comments, the examination must be evaluated to ensure that the proper ratios between sections are still satisfied. If deleting questions affects the content validity of the examination, the examination shall be withdrawn and a new examination administered.

(4)   The examiner grading the examination must incorporate into the master examination and into the answer key all changes in response to the facility licensee's comments and any corrections made while the examination was administered.

C. Grading the Examination

(1)   The written examination will not be graded until the facility licensee's comments have been resolved.

(2)   Normally, the examiner who wrote an examination will grade it. However, if the author is unavailable, the number of applicants is unusually large, or the CE wishes to expedite the grading process, he or she may have another examiner grade the examination.

(3)   The grader will indicate in red pen or pencil the number of points given to or deducted for each answer on the applicant's answer sheet.

(4)   If changing a grade is necessary, the grader or reviewer will do so by lining out the original grade so that it remains legible, briefly documenting the reason for the change on the applicant's answer sheet, and initialing the change. The grader or reviewer shall not use Wite-Out® and other methods that obscure the change.

(5)   After all the questions are graded, the grader will enter the test value, applicant's score, and applicant's grade in the *Results* section of the applicant's written examination cover sheet.

(6)   After grading the last applicant, the grader will review all of the answer sheets to ensure that all applicants were graded consistently.

(7)   The grader will fill in the *Written Examination Summary* section on page 1 of Form ES-301N-1, check the *Pass, Fail,* or *Waive* block, and sign and date the grader line in the *Examiner Recommendations* section. The grader should then forward the written examination package, including the master written examination and answer key, the applicants' examinations, and all associated forms (ES-301N-1 and ES-401N-1) to the CE. The CE will continue the process for the examination as outlined in ES-501N.

ATTACHMENTS/FORMS:

NONE

EXAMINER STANDARD 501N
POSTEXAMINATION ACTIVITIES

A. Purpose

This standard describes the procedures for assembling and reviewing the examination package and notifying the facility and applicants of the examination results.

B. Background

The goal is to complete licensing actions within 30 days of resolving the facility licensee's written examination comments.

C. Examination Report

The CE will prepare an examination report. The sample examination report included as Attachment 1 should be used as a guide.

(1)   The report documents the results of the examination, including generic strengths and weaknesses noted during administration of the operating tests, grading of the written examinations, and issues discussed at the exit meeting.

(2)   The report will also include a copy of the written examination with the answer key modified to incorporate facility comments.

(3)   Applicants' names and grades (i.e., Form ES-501N-1) will **not** be included in the examination report.

D. Results Package

The CE will also prepare an examination results letter which includes a public cover letter and its nonpublic enclosures. As a guide, the CE should use the sample examination results letter (Attachment 2 to this standard) and enclosures described below.

(1)   The enclosures that are nonpublic will include an examination grading summary sheet (Form ES-501N-1), individual operator examination reports (Form(s) ES-301N-1), and individual candidates' written examination answer sheets, if a written examination was administered.

E. Examination Reviews

(1)   The CE will forward the examination package to another examiner for peer review.

(2)   The peer reviewer will review the grading of the written examination and the operating test documentation to ensure that they meet the requirements of ES-403N and ES-303N, respectively. During these reviews, the peer reviewer should pay particular attention to the consistency of grading among all applicants tested.

(3)   After reviewing the package, the peer reviewer will check either the *Pass* or *Fail* block, then date and sign the peer reviewer line of each Form ES-301N-1. If the peer reviewer disagrees with the grading of either the written examination or the operating test, he or she will resolve the disagreement with the grader of that portion of the examination. The peer reviewer will return the package to the CE.

(4)   The CE will independently review the examination package then check either the *Issue* or *Deny* block and sign and date the *License Recommendation* section of each Form ES-301-1.

F. Licensing Action

(1)   The CE will give the examination package to the OLA who will prepare license and denial letters for all applicants tested. Attachments 3 and 4 are sample RO and SRO (conditional) license and denial letters. The license and denial letters will become an attachment to the results package.

(2)   The OLA will give the completed packages to the licensing official, who makes the final licensing decision. Per NRC policy, the licensing official is the BC/PD with program responsibility. A person acting for the BC/PD may **not** sign license or denial letters. The licensing official will consider all recommendations and sign each applicant's license or denial letter.

        Applicants who withdrew before taking any part of the license examination will not be sent a denial letter and will not be included in either package. Applicants who withdrew from the examination after it had started will be denied a license.

(3)   If the licensing official does not agree with the final recommendation, he or she will consult with the grader, peer reviewer, and CE to resolve the disagreement. Disagreements usually arise because of poor documentation. Therefore, it is very important for examiners to be accurate in grading and writing comments.

(4)   If the licensing official decides to overturn the recommendation, he or she will line out and initial the affected summary evaluations. The licensing official will enter a new summary evaluation in the appropriate block and explain the change in the *Operating Test Comments* section of Form ES-303N-1.

(5)   Once the licensing decisions are complete, the examiners should discard any marked documentation or rough notes for those applicants receiving licenses. Examiners should retain notes and documentation associated with proposed denials until the denial becomes final according to ES-502N.

(6)   For SRO-U applicants who fail an examination, the BC/PD should direct a case-specific review to decide if the applicant failed because of significant deficiencies in RO knowledge or abilities. As stated under 10 CFR 55.7, "Additional Requirements," "the Commission may, by rule, regulation, or order, impose upon any licensee such requirements, in addition to those established in the regulations in this part, it deems appropriate or necessary to protect health and to minimize danger to life and property." If the SRO-U applicant's deficiencies pose such a threat, the NRC may require the facility licensee to provide remedial training and reevaluation and submit evidence of its completion to the NRC.

G. Facility and Individual Notification

(1)   The CE may phone the facility staff to inform them of examination results only after the licensing official has reviewed and approved them.

(2)   After the Licensing Official has signed the license and denial letters, the OLA will send each applicant the following material with his or her letter:

- a copy of the "as given" Form ES-303N-1
- a copy of the applicant's written examination answer sheet

(3)   The secretary for research and test reactors personnel is responsible for distribution of the examination report package.

(4)   The OLA is responsible for distribution of the examination results package and the notifications to the individuals taking the examination per Item 2 above.

(5)   If any examinations are regraded in response to an applicant's request for review (refer to ES-502N), the original Form ES-501N-1 on file will be corrected by lining out the old grade, entering the new grade, and initialing the change.

H. Returning Facility Reference Material

If all applicants passed the examination, the CE should return the facility reference material when the licenses are issued. If any applicants failed the examination, the examiner should retain the reference material to support either resolution of appeals or preparation of retake examinations.

I. Record Retention

With the advent of ADAMS, the need to maintain hard (paper) copies of records is greatly reduced. The following documents should be maintained in ADAMS:

(1)   ES-201N, Attachment 1, "Corporate Notification Letter," with all enclosures (publicly available)

(2)   Form ES-201N-1, "Examination Assignment Sheet, with changes to identify the applicants that were actually examined (not publicly available)

(3)   ES-501N, Attachment 1, "Examination Report," with all enclosures (publicly available)

(4)   ES-501N, Attachment 2, "Examination Results Letter," cover letter only (publicly available)

(5)   Form ES-501N-1, "Examination Summary Sheet" (not publicly available)

(6)   Certificate Transmittal Cover Letter (not publicly available)

Each applicant's docket file should contain the original or a copy of the following items:

(1)   Form ES-303N-1 (all pages) with final, clean versions of page 5 and all associated documentation

(2)   all correspondence with the applicant

(3)   the applicant's written examination answer sheets

ATTACHMENTS/FORMS:

Attachment 1, Sample Examination Report (Public)
Attachment 2, Sample Results Cover Letter (Public)
Attachment 3, Sample License Letters (Nonpublic)
Attachment 4, Proposed Denial Letter (Nonpublic)
Form ES-501N-1, Research and Test Reactor Plant Examination, Results Summary Sheet (Nonpublic)

ATTACHMENT 1
SAMPLE EXAMINATION REPORT

[NRC Letterhead]

[Date]

[Applicant/Facility Name]
[Street Address]
[City, State Zip Code]

SUBJECT: INITIAL EXAMINATION REPORT NO. 50-[NUMBER]/OL-YY-NN, [FACILITY NAME]

Dear [Name]:

During the week of KEYBOARD(Enter date, then hit ALT-SHIFT-C), the U.S. Nuclear Regulatory
Commission (NRC) administered an initial operator licensing examination at your FIELD(facility) reactor.
The examination was conducted according to NUREG-1478, "Research and Test Reactor Operator
Licensing Examiner Standards," Revision 2, published in June 2007. Examination questions and
preliminary findings were discussed at the conclusion of the examination with those members of your
staff identified in the enclosed report.

In accordance with Title 10, Section 2.790, of the *Code of Federal Regulations*, a copy of this letter and
the enclosures will be available electronically for public inspection in the NRC Public Document Room or
from the Publicly Available Records (PARS) component of NRC's Agencywide Documents Access and
Management System (ADAMS). ADAMS is accessible from the NRC Web site at
http://www.nrc.gov/reading-rm/adams.html (the Public Electronic Reading Room). The NRC is
forwarding the individual grades to you in a separate letter which will not be released publicly. If you
have any questions concerning this examination, please contact [chief examiner] at [telephone number],
or via email at [initials]@nrc.gov.

Sincerely,

[Branch Chief with Program Responsibility]

Docket No. 50-[number]

Enclosures:
1. Examination Report No. 50-[number]/OL-YY-NN
2. Facility Comments with NRC Resolution
3. Corrected Written Examination

cc w/enclosures: Facility Training Manager
(Standard Distribution) plus [Appropriate Project Manager]

ATTACHMENT 1
SAMPLE EXAMINATION REPORT (Continued)

ENCLOSURE 1

EXAMINATION REPORT NO: 50-[NUMBER]/OL/YY-NN

[FACILITY]

Facility License No.:   [License No.]

Submitted by: _____   __/__/__
                    [Name], Chief Examiner   Date

Summary:

During the week of [date], the NRC administered licensing examinations to [number] SRO-I, [number] SRO-U, and [number] RO license applicants.   [Number] SROs and [number] ROs passed these examinations. All others failed.

Exit Meeting:
   Attendees:

              [Name, Title, Facility]
              [Name, Title, Facility]
              [Name], Examiner, NRC

At the conclusion of the site visit, the examiners met with representatives of the facility staff to discuss the results of the examinations. The examiners made the following observations concerning your training program:

a.   Areas of generic weaknesses were found in [brief statement of the areas of weakness identified during the exit meeting]. The facility committed to placing more emphasis in these areas in future training programs [Open Item number].

b.   Areas in which the examiners believe that the applicants exhibited good training and knowledge are [brief statement of any strengths identified during the exit meeting].

c.   The facility presented comments on the written examination, and these comments have been incorporated into the examination attached to this report.

OR

c.   The facility presented comments on the written examination, and these comments along with their resolutions are included as Attachment 2 to this report.

ATTACHMENT 1
SAMPLE EXAMINATION REPORT (Continued)

ENCLOSURE 2

FACILITY COMMENTS WITH NRC RESOLUTION

Question 4.01

Comment: Control Room Log Book should be included as an additional correct response.
Justification: Administrative Procedure AP-4.0, page 8.

Question 4.01

NRC Resolution: Comment accepted. The alternative answer is acceptable but not required for a
                complete answer.

ATTACHMENT 1
SAMPLE EXAMINATION REPORT (Continued)

ENCLOSURE 3

CORRECTED WRITTEN EXAMINATION

ATTACHMENT 2
SAMPLE RESULTS COVER LETTER

[NRC Letterhead]
[Date]

[Applicant/Facility Name]
[Street Address]
[City, State Zip Code]

SUBJECT: INITIAL EXAMINATION RESULTS LETTER NO. 50-[NUMBER]/OL-YY-NN, [FACILITY
             NAME]

Dear [Name]:

This letter forwards to you personal privacy information associated with the operator licensing
examinations administered during the week of [date]. This information supplements the publicly
available information in the examination report. Enclosure 1 lists individual grades for all candidates.
Enclosure 2 includes individual candidate summaries for all candidates (license/denial letters and
supporting documentation).

If you have any questions regarding the examination, please contact [chief examiner] at
(301)415-[number], or via email at [initials]@nrc.gov.

                                        Sincerely,

                                        [Branch Chief with Program Responsibility]

Docket No. 50-[number]

Enclosures:
As Stated

DISTRIBUTION:
PUBLIC (w/o enclosures)
Facility File ([LA]) O6-D17

ADAMS ACCESSION #: ML011   TEMPLATE #:NTR-079

OFFICE	branch/program:CE	E	branch/program:OLA		branch/program:BC	
NAME	[EXAMINER]:[SECRETARY]		[OLA]		[BC/PD]	
DATE	[DATE]		[DATE]		[DATE]	

C = COVER   E = COVER & ENCLOSURE   N = NO COPY
                                    OFFICIAL RECORD COPY

ATTACHMENT 3
SAMPLE LICENSE LETTERS

[NRC Letterhead]

[Date]

[LICENSE]

Docket No. 55-[number]

[Applicant/Facility Name]
[Street Address]
[City, State Zip Code]

Dear [Name]:

Pursuant to the Atomic Energy Act of 1954, as amended, the Energy Reorganization Act of 1974, as amended (Public Law 93-438), and subject to the conditions and limitations incorporated herein, the U.S. Nuclear Regulatory Commission hereby licenses you to manipulate all controls of the [name of facility, facility license number].

Your License No. is OP-[number]. Your Docket No. is 55-[number]. The effective date is [date]. Unless sooner terminated, renewed, or upgraded, this license shall expire 6 years from the effective date.

This license is subject to the provisions of Title 10, Section 55.53, "Conditions of Licenses," of the *Code of Federal Regulations*, with the same force and effect as if fully set forth herein.

While performing licensed duties, you shall observe the operating procedures and other conditions specified in the facility license which authorizes operation of the facility.

The issuance of this license is based on examination of your qualifications, including the representations and information contained in your application for this license.

A copy of this license has been made available to the facility licensee.

For the Nuclear Regulatory Commission

[Name and title of licensing official]

cc: [Facility representative who signed the applicant's NRC Form 398]

ATTACHMENT 3
SAMPLE LICENSE LETTERS (Continued)

[NRC Letterhead]
[Date]

[LICENSE]

Docket No. 55-[number]

[Applicant Name]
[Street Address]
[City, State Zip Code]

Dear [Name]:

Pursuant to the Atomic Energy Act of 1954, as amended, the Energy Reorganization Act of 1974, as amended (Public Law 93-438), and subject to the conditions and limitations incorporated herein, the U.S. Nuclear Regulatory Commission hereby licenses you to direct the licensed activities of licensed operators at, and to manipulate all controls of, the [name of facility, facility license number].

Your License No. is SOP-[number]. Your Docket No. is 55-[number]. The effective date is [date]. Unless sooner terminated or renewed, this license shall expire 6 years from the effective date.

This license is subject to the provisions of Title 10, Section 55.53, "Conditions of Licenses," of the *Code of Federal Regulations*, with the same force and effect as if fully set forth herein.

While performing licensed duties, you shall observe the operating procedures and other conditions specified in the facility license which authorizes operation of the facility and shall comply with the following condition:

   You shall wear corrective lenses while performing the activities for which you are licensed.

The issuance of this license is based on examination of your qualifications, including the representations and information contained in your application for this license.

A copy of this license has been made available to the facility licensee.

                    For the Nuclear Regulatory Commission

                    [Name and title of licensing official]

cc: [Facility representative who signed the applicant's NRC Form 398]

ATTACHMENT 4
PROPOSED DENIAL LETTER

[NRC Letterhead]
[Date]

[Applicant/Facility Name]
[Street Address]
[City, State Zip Code]

Dear [Name]:

This letter is to inform you that your grade on the [operating test, written examination, or both] taken on [date(s)], in connection with your application for a [reactor operator, senior reactor operator] license for the [facility name], indicates that you did not pass that [test, examination, or both]. As a result, the U.S. Nuclear Regulatory Commission (NRC) proposes to deny your application. Enclosed is a copy of the [operating test, written examination, or both] results indicating those areas in which you exhibited deficiencies. [A copy of the master answer key is also provided.]

If you accept the proposed denial and decline to request either an informal NRC staff review or a hearing within 20 days as discussed below, this proposed denial will become a final denial. You may then reapply for a license according to Title 10, Section 55.35, "Re-applications," of the *Code of Federal Regulations* (10 CFR 55.35) subject to the following conditions:

*   a.   Because you passed [a written examination, an operating test] on [date], you may request a waiver of that portion. This waiver will be granted by the NRC and will be valid up to 1 year from your [examination, test] date.

*   b.   Because you did not pass the [written examination, operating test] administered to you on [date], you will be required to retake [a written examination, an operating test].

*   c.   You may reapply for a license 2 months from the date of this letter.

OR

** a.   Because this is a [second, subsequent] examination failure, you will be required to retake the written examination and the operating test.

** b.   You may reapply for a license [6, 24] months from the date of this letter.

OR

*** a.   Because you passed neither the written examination nor the operating test administered to you on [date(s)], you will be required to retake both the operating test and the written examination.

*** b.   You may reapply for a license [2, 6, 24] months from the date of this letter.

If you do not accept the proposed denial, you may, within 20 days of the date of this letter, take one of the following actions:

ATTACHMENT 4
PROPOSED DENIAL LETTER (Continued)

2

- You may request an informal NRC staff review of the grading of your examination. Your written request must be sent to the [appropriate division director], U.S. Nuclear Regulatory Commission, Washington, DC 20555. Your request must identify the portions of your examination that you believe were graded incorrectly or too severely. In addition, you must provide the basis, including supporting documentation (such as procedures, instructions, computer printouts, and chart traces), in as much detail as possible, to support your contention that certain of your responses were graded incorrectly or too severely.

  The NRC will review your contentions, reconsider your grading, and inform you of the results. If the proposed denial is sustained, you will have the opportunity to request a hearing pursuant to 10 CFR 2.103(b)(2) at that time.

- You may request a hearing pursuant to 10 CFR 2.103(b)(2). Submit your request, in writing, to the Secretary, U.S. Nuclear Regulatory Commission, Washington, DC 20555-0001, Attention: Rulemakings and Adjudications Staff, with a copy to the Assistant General Counsel for Hearings, Enforcement and Administration, Office of the General Counsel, at the same address. (Refer to 10 CFR 2.302, "Filing of Documents," for additional filing options and instructions.)

Pursuant to 10 CFR 55.35, you may not reapply for a license until your license has been finally denied. Failure on your part to exercise one of these options within 20 days constitutes a waiver of your opportunity for informal review and your right to demand a hearing and, for the purpose of reapplication under 10 CFR 55.35, renders this letter a notice of final denial of your application, effective as of the date of this letter.

If you have any questions, please contact [name] at [telephone number].

Sincerely,

[Name and title of licensing official]

Docket No. 55-[number]

Enclosures:
As stated

cc: [Facility representative who signed the applicant's NRC Form 398]

CERTIFIED MAIL—RETURN RECEIPT REQUESTED

---

*   Use for initial reactor operator or senior reactor operator license applicants who passed either the written examination or the operating test but failed the other.
**   Use for second and subsequent retake applicants.
***   Use for applicants who failed both the written examination and the operating test.

FORM ES-501N-1
RESEARCH AND TEST REACTOR PLANT EXAMINATION

Results Summary Sheet

Personally Identifiable Information—
Withhold under 10 CFR 2.390

RESEARCH AND TEST REACTOR EXAMINATION RESULTS SUMMARY SHEET

Facility:

Written Exam Date:

Examiners:

Plant Status:

Operating Exam Dates:

OVERALL RESULTS	TOTAL No.	No. Passed	% Passed	No. Failed	% Failed	HOT
Reactor Operator						
Senior Operator						

Name	Docket 55-	Type[1]	Written Grade				Results[2]/Initials[3]	
			A	B	C	Total	Written	Operating

NOTES:
[1] 1=RO, 2=SRO-I, 3=SRO-U, 4=RO retake, 5=SRO-I retake, 6=SRO-U retake, 7=SRO-Fuel Handler
[2] P=Passed, F=Failed, W=Waived, N/A=Not Applicable
[3] Enter the initial of the examiner who graded the written examination or administered the operating test.

EXAMINER STANDARD 502N
ADMINISTRATIVE REVIEWS AND HEARINGS

A. Purpose

This standard specifies the NRC's policies and practices for processing informal staff reviews of license application denials, issuing final denials of license applications, and reapplying for a license after a final denial of an application has been issued.

B. Background

The letter sent to applicants who fail an operator licensing examination in ES-501 constitutes a proposed license denial. The letter informs the applicant of his or her options for appeal as specified in Section C below and includes a list of the deficiencies noted.

Applicants who fail an operator licensing examination may reapply pursuant to the provisions of 10 CFR 55.35, only after they have been issued a final denial of their application, and only if they have not requested either an informal NRC review or a hearing. The NRC will not accept a reapplication while a request for either an informal NRC review or a hearing is pending.

The procedures that apply when the NRC has denied an application for an operator license because the applicant fails to meet the eligibility requirements in 10 CFR 55.31 are similar to those for processing informal NRC reviews and license denials. The details of this process are described in Section E below.

C. Applicant Responsibilities

(1)   An applicant who does not appear to meet the experience and training requirements for a license may be asked to provide additional information in accordance with ES-202N, "Eligibility Requirements and Guidelines." If the NRC still denies the application after the applicant provides the additional information requested by the NRC, the applicant may exercise one of the following options within 20 days after the date of the proposed denial letter:

   a.   Do nothing. The proposed denial letter then becomes the final denial. The applicant may reapply after obtaining the requisite training or experience.

   b.   Request reconsideration of the application denial. Applicants must submit such requests to the Director, Division of Regulatory Improvement Program, Office of Nuclear Reactor Regulation, U.S. Nuclear Regulatory Commission, Washington, DC 20555-0001. The applicant's submittal must clearly state the basis for the request.

   c.   Request a hearing pursuant to 10 CFR 2.103(b)(2). Applicants must submit such requests to the Office of the Secretary, U.S. Nuclear Regulatory Commission, Washington, DC 20555-0001, Attention: Rulemakings and Adjudications Staff, with a copy to the Associate General Counsel for Hearings, Enforcement, and Administration, Office of the General Counsel, at the same address. (Refer to 10 CFR 2.302, "Filing of Documents," for additional filing options and instructions.)

(2)   If an applicant fails the operator licensing written examination or operating test (or both) and receives a proposed license denial letter issued in accordance with ES-501N, the applicant has 20 days from the date of the letter to exercise one of the following three options:

a. Do nothing. The proposed denial letter then becomes the final denial. The applicant may reapply, pursuant to 10 CFR 55.35, 2 months after the date of the first denial letter, 6 months after the second denial, and 24 months after each successive denial.

b. Request that the NRC administratively regrade the written examination, the operating test, or both, in light of new information to be provided by the applicant. Applicants must submit such requests to the Director, Division of Regulatory Improvement Program, Office of Nuclear Reactor Regulation, U.S. Nuclear Regulatory Commission, Washington, DC 20555-0001. If the applicant submits such a request, the NRC will not consider a reapplication pursuant to 10 CFR 55.35 until a denial is final.

The applicant's request for administrative review must identify the item(s) for which additional review is requested and must include documentation supporting the item(s) in contention. The applicant is responsible for ensuring that the request and the supporting documentation are sent to RTROL within 20 days after the date of the proposed denial letter.

If the NRC staff administratively reviews a failure and determines that the applicant did not provide sufficient basis to justify passing grades on all sections of the licensing examination, the NRC will issue a letter to the applicant sustaining the proposed denial. The applicant may then request a hearing pursuant to 10 CFR 2.103(b)(2). In such instances, the applicant must submit a request for a hearing after an administrative review within 20 days after the date on the letter from the RTROL sustaining the proposed denial. In addition, the applicant must submit the hearing request in accordance with Section C.2(c), below.

If the applicant does not request a hearing when the RTROL sustains the proposed denial, the proposed denial becomes the final denial. The applicant may then reapply for a license, pursuant to 10 CFR 55.35, 2 months after the date of the first sustained denial letter, 6 months after the second denial, and 24 months after each successive denial.

c. Request a hearing as provided by 10 CFR 2.103(b)(2). The applicant must submit the hearing request to the Office of the Secretary, U.S. Nuclear Regulatory Commission, Washington, DC 20555-0001, Attention: Rulemakings and Adjudications Staff, with a copy to the Associate General Counsel for Hearings, Enforcement, and Administration, Office of the General Counsel, at the same address. (Refer to 10 CFR 2.302 for additional filing options and instructions.) If the applicant requests a hearing, the NRC will not consider a reapplication pursuant to 10 CFR 55.35 until the denial is final.

D. Facility Licensee Responsibilities

The NRC may ask the facility licensee to provide reference materials, technical support, and a confirmation of the validity of the test items, as necessary for the NRC staff to evaluate and resolve any concerns raised by a license applicant who asks the NRC to reconsider a proposed denial of an application or license.

E. NRC Responsibilities

(1) The NRC will conduct administrative reviews of 10 CFR Part 55 license application denials based on eligibility as described in Section F.1, below.

(2)   The NRC will conduct administrative reviews of 10 CFR Part 55 license denials based on examination failures as described in Section F.2, below.

(3)   The NRC will conduct 10 CFR Part 55 operator licensing hearings in accordance with 10 CFR Part 2

F. Administrative Review Procedures

1. Application Denial

If an applicant requests an administrative review in accordance with Section C.1.b, the RTROL will generally complete its review of the applicant's eligibility within 60 days of receiving the request. Upon completing its review, the RTROL will notify the applicant in writing as to whether he or she will be allowed to retake the license examination. If the review leads the RTROL to sustain the original denial, the applicant may request a hearing pursuant to 10 CFR 2.103(b)(2).

2. Examination Results

If an applicant requests an administrative review in accordance with Section C.2.b, the RTROL will generally complete its review, as follows, within 45 days after receiving the request.

a.   The RTROL will determine whether to (1) review the appeal internally or (2) convene a three-person board to review the applicant's documented contentions. The appeal board will normally be composed of a BC/PD and two examiners or subject matter experts, but no one who was involved with the applicant's licensing examination.

For written examinations, the review will generally focus only on those questions that the applicant is contesting. The review shall evaluate the original grading of the applicant's (or applicants') examination(s), the reference material supplied by the facility licensee, and the contentions and supporting documentation provided by the applicant(s). If there are multiple appeals, all question deletions and answer key changes will be applied equally to each appellant's examination, without regard to who submitted the complaint. Moreover, in those rare instances when a generic finding results in a change to an answer key (e.g., failure to provide a print or other reference necessary to answer a question), the corrective action may be applied, as appropriate, to adjust the grading of other questions that were not contested.

For operating tests, the review shall evaluate the examiner's comments, the examination report, the test that was administered, and the contentions and supporting documentation provided by the applicant or facility licensee (e.g., plant system descriptions, operating procedures, logs, chart recorder traces, and process computer printouts).

b.   Based on the findings and recommendations from the review, the RTROL will decide whether to sustain or overturn the applicant's license examination failure. The RTROL will then notify the applicant in writing of the results of the review.

c.   If the proposed denial was overturned, then the RTROL will review the examination results of the other applicants to determine whether any of the licensing decisions are affected and update the master examination file to reflect any test item deletions or answer key changes.

G. Notes

(1)   A BC/PD or above will sign all letters informing an applicant of an examination failure or an application denial. The BC/PD will include the appropriate licensing assistant on distribution for all correspondence generated according to this standard.

(2)   All facility licensee representatives who signed the license applications will be sent copies of all external correspondence generated because of this standard.

(3)   The BC/PD is responsible for informing management of review requests from license applicants.

(4)   All correspondence referenced in this standard will be sent to the applicant via certified mail, with return receipt requested.

ATTACHMENTS/FORMS:

Attachment 1, Sample License Notification Letter from Division Director
Attachment 2, Proposed License Denial Letter from Division Director
Attachment 3, Initial Application Denial Letter from Division Director

ATTACHMENT 1
SAMPLE LICENSE NOTIFICATION LETTER

[NRC Letterhead]

[Date]

[Applicant/Facility Name]
[Street Address]
[City, State Zip Code]

Dear [Name]:

As a result of the additional information you supplied in your letter of [date], we have decided that you passed the [written examination, operating test] administered to you on [date(s)], and satisfy the requirements of Title 10, Section 55.33(a), of the *Code of Federal Regulations* (10 CFR 55.33(a)) for approval of your license application.

The NRC will issue you a [reactor operator or senior reactor operator] license, pursuant to 10 CFR 55.51, "Issuance of Licenses," in a separate letter.

{For your information, I am enclosing a copy of the staff's resolution of each of your comments.} If you have any questions, please contact [name] at [telephone number].

Sincerely,

[Division Director with Program Responsibility]
Office of Nuclear Reactor Regulation

Docket No. 55-[number]

{Enclosure:}
{As stated}

{ }   Include only if letter does not discuss comment resolutions.

ATTACHMENT 2
PROPOSED LICENSE DENIAL LETTER

[NRC Letterhead]
[Date]

[Applicant/Facility Name]
[Street Address]
[City, State Zip Code]

Dear [Name]:

Using the information you supplied in your letter of [date], we have reviewed the grading of the [written examination, operating test] administered to you on [date(s)]. We find that you did not pass the [examination, test]. Consequently, the proposed denial of your license application is sustained. We have enclosed our review results with this letter.

If you do not request a hearing within 20 days of the date of this letter, the proposed denial will become a final denial, and you may reapply for a license according to Title 10, Section 55.35, "Re-applications," of the *Code of Federal Regulations* (10 CFR 55.35) subject to the following conditions:

* a.   You may request a waiver of the portions of the examination that you passed. The NRC will grant this waiver provided that you submit it within 1 year from your previous examination date.

* b.   You are required to take an operator licensing examination, consisting of the portions of the examination not waived.

* c.   You may reapply for a license 2 months from the date of this letter. The NRC will schedule an examination upon request by you or your facility management.

OR

** a.   Because this is a [second, subsequent] examination failure, you will be required to retake all portions of the written examination and the operating test.

** b.   You may reapply for a license [6, 24] months from the date of this letter. The NRC will schedule an examination upon request by you or your facility management.

If you do not accept the proposed denial, you may, within 20 days of the date of this letter, request a hearing according to 10 CFR 2.103(b)(2). Submit your request, in writing, to the Secretary of the Commission, U.S. Nuclear Regulatory Commission, Washington, DC 20555, with a copy to the Assistant General Counsel for Hearings, Office of the General Counsel, at the same address.

Failure on your part to request a hearing within 20 days constitutes a waiver of your right to demand a hearing and, for the purpose of reapplication under 10 CFR 55.35, renders this letter a notice of final denial of your application, effective as of the date of this letter.

If you have any questions, please contact [name] at [telephone number].

                              Sincerely,

                              [Division Director with Program Responsibility]
                              Office of Nuclear Reactor Regulation

Docket No. 55-[number]

Enclosure: As stated

cc: [Facility representative who signed the applicant's NRC Form 398]

CERTIFIED MAIL—RETURN RECEIPT REQUESTED

---

*   Use for initial reactor operator or senior reactor operator license applicants who passed either the written examination or the operating
        test but failed the other.
**   Use for second and subsequent retake applicants.

ATTACHMENT 3
INITIAL APPLICATION DENIAL LETTER

[NRC Letterhead]
[Date]

Docket No. 55-[number]

[Applicant/Facility Name]
[Street Address]
[City, State Zip Code]

Dear [Name]:

In response to your letter dated [date], we have reviewed the denial of your application for a [reactor operator, senior reactor operator] license issued on [date]. Our review of your application indicates that you still do not meet the eligibility requirements.

[Discussion of deficiencies with reference to relevant parts of 10 CFR 55.31, ES-202N, or NRC-approved facility training program] When you have met the requirements of Title 10, Section 55.31, "How To Apply," of the *Code of Federal Regulations* (10 CFR 55.31), you may submit another application.

If you do not accept this denial, you may, within 20 days of the date of this letter, request a hearing pursuant to 10 CFR 2.103(b)(2). Submit your request, in writing, to the Secretary of the Commission, U.S. Nuclear Regulatory Commission, Washington, DC 20555, with a copy to the Assistant General Counsel for Hearings, Office of the General Counsel, at the same address.

If you have any questions, please contact [name] at [telephone number].

Sincerely,

[Division Director with Program Responsibility]
Office of Nuclear Reactor Regulation

cc: [Facility representative who signed the applicant's NRC Form 398]

CERTIFIED MAIL—RETURN RECEIPT REQUESTED

EXAMINER STANDARD 601N
REQUALIFICATION EXAMINATION ADMINISTRATION

A. Purpose

This standard describes the methods to be used to prepare, conduct, and grade requalification examinations meeting the provisions of 10 CFR 55.59(a)(2)(iii) and 10 CFR 55.59(c)(7). This standard also describes the methods to be used to evaluate the facility licensee's requalification program. The program will be applied to licensees where the NRC has lost confidence in the licensee's ability to conduct its own examination. Situations that could result in a **for cause** requalification examination are significant operator errors, poor requalification inspection results, and allegations of significant training program deficiencies.

B. General

When possible, a team of both NRC and facility members will develop requalification examinations. The examination will be based on the NRC-approved requalification program. Simultaneous evaluation of operator performance by the facility and the NRC will allow the NRC to assess both individual and program performance. The examination will consist of a written examination and an operating test.

C. Examination Schedule

(1)   At least 90 days before examination administration, the NRC will notify the facility licensee of the decision to administer an NRC requalification examination and ask whether the facility has the resources to prepare the written examination, the operating tests, or both. If the facility commits to preparing any part of the examination, the NRC will schedule a site visit to explain the examination process and answer any questions that the facility may have. The NRC will also send the facility a letter in the format of the Notification Letter in Attachment 1 to this standard.

(2)   At least 60 days before examination administration, the facility licensee will send the NRC reference material (including its examination bank, if available). The NRC will evaluate the material for adequacy. The facility will identify the following:

- systems covered during the requalification cycle
- new or recently modified systems
- fuel accountability and handling
- building and prestartup requirements
- experiment handling
- radiation monitoring and control
- other systems and procedures required for the safe operation of the facility

The facility will also provide a list of operators to be examined and the names of any staff (referred to as "facility staff") assigned to the examination team. The facility staff will provide technical assistance in the development and review of the written examination questions and the operating test. Normally, the facility staff will act as the facility evaluators during the operating tests.

(3)   At least 30 days before the examination administration date, the facility should provide any portions of the requalification examination that it has committed to preparing. The NRC will evaluate the material and any facility-prepared examinations for adequacy. Also, the CE will reconfirm the schedule and recommend any changes necessary. The NRC reserves the right to reschedule the examination if it finds the facility-supplied materials or examinations inadequate.

D. Examination Preparation

1. Written Examination Preparation

The written examination will be an open-reference, three-section examination and will not discriminate between RO and SRO knowledge levels.

Section A tests operators' knowledge of reactor theory, thermodynamics, and facility operating characteristics. Section B tests operators' knowledge of normal, abnormal, and emergency procedures, and administrative controls (including technical specifications, the emergency plan, administrative procedures, and radiological procedures). Section C tests operators' knowledge of facility systems, with the focus on systems applicable to reactor operations (i.e., process and radiological monitoring).

The author should prepare 20 questions per section. Questions should come from the facility's requalification examination bank, with appropriate references. The questions should be objective (i.e., multiple choice or matching).

2. Written Examination Reviews

The entire examination team will review the written examination evaluating each question for appropriateness, technical accuracy, and clarity. The examination team must repeat this evaluation for any changes made as a result of this review.

3. Operating Test Preparation

The operating test is also open reference. Each operator will be evaluated on his or her ability to complete a minimum of five tasks satisfactorily. The operating tests will differentiate between RO and SRO knowledge levels and abilities. An RO is only responsible for RO tasks; an SRO is responsible for all tasks.

The preparer will specify the criteria for satisfactory completion of each task indicating "critical steps." A critical step is one that, if performed incorrectly or not at all, would prevent the system from operating safely or prevent completion of an essential safety action. Examples of essential safety actions include the following:

-   effective manipulation of the controls affecting reactivity
-   actuation of a reactor trip
-   compliance with technical specifications
-   reduction in excessive levels of radiation and guarding against personnel exposure

When preparing tasks, use of a copy of the procedure (or sections) marking steps considered critical is acceptable. For tasks with no written procedure, the preparer will supply a written description of the task, list of expected actions, and indication of critical steps.

Each operator must be tested on at least four different types of tasks (listed below). The examiner may design the tasks to flow from one to the next, such as having the operator perform a startup as the first task, then while critical, simulate an instrument failure as the second task.

- **Reactivity task:** Tasks that have the operator perform a startup, significant power change, or shutdown.

- **Abnormal event:** Tasks that have the operator respond to an instrument failure, component failure, radiation monitor alarm, or similar problem.

- **Emergency event:** Tasks that have the operator respond to events such as a building evacuation, a large reactor pool leak, high-radiation levels.

- **Outside event:** Tasks that have the operator perform duties outside the control room, such as refueling or surveillance on local radiation monitors.

Additional tasks will be developed at the discretion of the CE.

Each task will include at least two questions to be asked at the completion of the task. Questions may be selected from facility question banks or from previously administered NRC examinations. To the maximum extent possible, the questions should have the following characteristics:

- be based on the task or system being operated
- discriminate between RO/SRO responsibilities, where appropriate
- emphasize knowledge required for task performance or procedure use and compliance

4. Operating Test Review

All team members must agree on critical steps. The CE will resolve any disagreements in the selection of tasks.

NRC examiner(s) will evaluate the facility-identified systems, tasks, and questions to ensure that the examination tests knowledge and abilities appropriate for operators at that facility.

The NRC may substitute any or all of the facility-provided tasks and questions with ones selected or developed by the NRC. However, facility-developed questions or tasks should be used unless they fail to meet NRC standards. The NRC CE has the final authority in deciding examination content.

E. Examination Administration

1. Written Examination

Either an NRC examiner or a facility staff member will proctor the written examination. The proctor will brief the operators using Attachment 1, Enclosure 3. Operators have 3 hours to complete the written examination.

2. Operating Test

Efforts must be made to avoid compromising the operating test. Operators examined on the same day may all perform the same tasks. Operators examined on subsequent days must perform a minimum of two tasks that were not previously administered, or the operators must complete a postexamination security agreement. The CE may agree to other reasonable precautions.

The preferred method for administering operating tests is to have the facility examination team member(s) act as the facility evaluator(s). The facility evaluator(s) should conduct the examination with the NRC examiner observing. If the facility cannot supply a facility evaluator, the NRC examiner will conduct the examination. NRC examiners may ask followup questions directly of the operator after task completion. However, to the extent possible, these questions will be asked through the facility evaluator.

The facility evaluator will brief the operator, using the "Briefing Checklist" (Attachment 2). If desired, the operators may be briefed as a group before the start of the operating tests.

Immediately after the operator completes a task, the facility evaluator will ask the two pre-scripted questions associated with the task. The evaluator may ask additional questions for clarification or verification of the tasks performed. The questions should not expand the scope of the pre-scripted questions.

Any additional questions asked should be reviewed with the facility evaluator as soon as possible after the tasks are completed.

Normally, the evaluator will observe task performance passively. The evaluator should not ask questions or present the next task while the operator is making console control manipulations. New tasks should be presented when the facility is in a steady-state condition.

The NRC examiner will ensure that the facility evaluator conducts an appropriate examination. If the NRC examiner determines that the examination is inadequate to make a pass/fail determination, the NRC examiner will discuss this concern with the facility evaluator. If the evaluator's examination continues to be unsatisfactory, the NRC examiner may choose to conduct the remainder of the examination with the facility evaluator grading in parallel. If this option is chosen, the CE will be informed at the completion of the individual examination. As soon as possible after such issues arise, the NRC CE will resolve with facility representatives all unforeseen technical questions or issues that could result in an operator failing the examination.

F. Examination Grading

1. Written Examination

The facility and NRC evaluators will grade the examinations independently. Grading should be completed within 5 workdays of examination administration. They will record the grades on the "Examination Cover Sheet," Form ES-601N-2.

Each operator must achieve a score of at least 70 percent overall as graded by the NRC to pass the examination. A score of less than 70 percent on one section is not a failure but indicates an area requiring remedial action by the facility.

2. Operating Tests

a.   Deficiencies will be recorded using the "Requalification Examination Operating Test Record," Form ES-601N-1. Typically, significant deficiencies will be associated with critical steps. Additional critical steps for a task may be identified because of unanticipated operator actions. These will be identified and conflicts resolved in the same manner as the original critical steps (see paragraph D.4).

b.   If an operator incorrectly performs or fails to perform a critical step, the task may be graded as unsatisfactory if the deficiency jeopardizes the safety of the facility or has a significant safety impact on the public. Failure to perform, or incorrect performance of, two or more critical steps will result in failure of the task.

c.   Each operator must successfully complete 70 percent of the tasks and correctly answer 70 percent of the prepared task questions to pass the test.

d.   The NRC will notify the facility immediately of any operator whose performance requires immediate removal from licensed duties.

G. Requalification Program Evaluation

The NRC will evaluate the program for any deficiencies or weaknesses based on examination and test adequacy and operator performance. The wide variance in staff sizes and the small number of operators that will typically be examined precludes performing a statistical program evaluation. However, if the NRC notes any of the following items during the examinations, it will consider a subsequent evaluation of the facility's training program:

(1)   More than 50 percent of the operators fail the examination (four or more operators are evaluated).

(2)   A significant deviation in pass/fail results exists between the NRC and the facility (i.e., the facility is less conservative in the evaluation of more than one operator).

(3)   The facility evaluator is unable to administer a satisfactory examination (e.g., leads the operator, cues by providing answers, or performs steps for the operator).

H. Actions for Requalification Program Deficiencies

The NRC will consider the following actions for any program deficiencies:

(1) request that the facility review program deficiencies and identify corrective actions to improve operator performance

(2) meet with senior facility management to review program deficiencies, determine root causes, propose corrective actions, and develop a schedule for implementing corrective action and followup inspections and examinations

(3) determine the following:

    a.   the significance of generic deficiencies identified during the program evaluation
    b.   how recent facility events relate to licensed operator performance

(4) develop NRC recommendations

Additional actions may be taken at the discretion of the Director of the Office of Nuclear Reactor Regulation or his designee.

I. Final Requalification Program Evaluation Report

A final requalification results summary sheet (Form ES-601N-3) will be prepared when the grading of requalification examinations has been completed. A complete copy of the report will be filed in the facility examination file.

J. Individual Requalification Examination Report

Form ES-601N-4 is filled out for each operator who has received an NRC requalification examination. This report (original) is filed in the individual's docket folder in the licensing assistant's office, with a copy sent to the facility and a copy filed in the facility examination file.

K. Record Retention

(1) When the requalification evaluation has been completed, the facility will receive a copy of all NRC-administered written and operating examinations.

(2) Material relating to an individual failure will be retained as necessary to support denial of license renewal per 10 CFR 55.57(b)(2)(iii). This includes the following:

    a.   examination cover page for all examinations
    b.   the portions of the examination that resulted in failure

(3) Examination Security

    To ensure examination security, the facility staff will be asked to follow examination security restrictions. These restrictions begin with the first review of the examination and continue until the examination is concluded. Each facility staff member will be required to sign the

"Preexamination Security Agreement" (Attachment 1, Enclosure 4) before reviewing the actual examination and the "Postexamination Security Agreement" at the conclusion of the examination process (Attachment 1, Enclosure 4).

ATTACHMENTS/FORMS:

Attachment 1, Notification Letter
Attachment 2, Briefing Checklist—Operating Test Tasks
Form ES-601N-1, Requalification Examination Operating Test Record
Form ES-601N-2, Examination Cover Sheet
Form ES-601N-3, Research and Test Reactor Requalification Results Summary Sheet
Form ES-601N-4, Individual Requalification Examination Report

ATTACHMENT 1
NOTIFICATION LETTER

[NRC Letterhead]
[Date]

[Applicant/Facility Name]
[Street Address]
[City, State Zip Code]

SUBJECT: CORPORATE NOTIFICATION LETTER, 50-[NUMBER]/OL-YY-NN, [FACILITY NAME]

Dear [Name]:

This letter confirms arrangements [I or examiner name] have made with [you or facility contact] for the administration of requalification examinations at the [facility]. The written and operating examination, scheduled for the week of [date], will be performed according to Examiner Standard 601N in the "Operator Licensing Standards for Research and Test Reactors," (NUREG-1478), issued in June 2007. You should have a copy of this standard.

Please furnish the material listed in Enclosure 1, "Reference Material Requirements," at least 60 days before the examination date to the following address:

(U.S. Postal Service)	(Private Mail Carrier)
ATTN: [Chief Examiner]	ATTN: [Chief Examiner]
[Appropriate Mail Stop]	[Appropriate Mail Stop]
U.S. Nuclear Regulatory Commission	U.S. Nuclear Regulatory Commission
Washington, DC 20555-0001	11555 Rockville Pike
	Rockville, MD 20852-2738

Failure to supply the reference material may result in a postponement of the examination. You may volunteer to submit a written examination for use during the examination week in addition to the material requirements of Enclosure 1. Submission of a written examination is optional. However, if you do submit a written examination, those personnel who developed it will be subject to the security restrictions described below.

The facility should provide at least one employee to complete the examination team. These employees should be licensed or previously licensed senior reactor operators at your facility or a similar facility. These individuals must not be scheduled for an examination administered by the U.S. Nuclear Regulatory Commission (NRC) during this visit. These facility representative(s) must sign a security agreement (Enclosure 4).

The facility representative(s) may continue to train operators with the understanding that they will not describe details of the examination, either in scope or content. Should questions arise that are on the examination, the facility representatives may answer them provided that they give no indication that the question is on the examination.

You are responsible for providing adequate accommodations to properly conduct the examinations. Enclosure 2, "Administration of Requalification Examinations," describes NRC requirements for conducting the examinations. Enclosure 3 contains the "NRC Rules and Guidance for Examinees" that will be in effect during the administration of the written examination. You are responsible for ensuring that all operators are aware of these rules.

This letter contains information collection requirements that are subject to the Paperwork Reduction Act of 1995 (44 U.S.C. 3501 et seq.). These information collections were approved by the Office of Management and Budget, approval number 3150-0018.

The burden to the public for these mandatory information collections is estimated to average 7.7 hours per response, including the time for reviewing instructions, searching existing data sources, gathering and maintaining the data needed, and completing and reviewing the information collection. Send comments regarding this burden estimate or any other aspect of these information collections, including suggestions for reducing the burden, to the Records and FOIA/Privacy Services Branch (T-5 F52), U.S. Nuclear Regulatory Commission, Washington, DC 20555-0001, or by Internet electronic mail to INFOCOLLECTS@NRC.GOV; and to the Desk Officer, Office of Information and Regulatory Affairs, NEOB-10202, (3150-0018), Office of Management and Budget, Washington, DC 20503.

The NRC may not conduct or sponsor, and a person is not required to respond to, a request for information or an information collection requirement unless the requesting document displays a currently valid OMB control number.

If you have any questions on the evaluation process, please contact me at [telephone], or email at [initials@nrc.gov]

Sincerely,

[Appropriate Branch Chief or Program Director or Above]
[Appropriate Branch]
[Appropriate Division]
Office of Nuclear Reactor Regulation

Docket No. 50-[number]

Enclosures:
1.   Reference Material Requirements
2.   Administration of Requalification Examinations
3.   NRC Rules and Guidance for Examinees
4.   Security Agreements

cc w/enclosures:
[Name], Reactor Supervisor

ENCLOSURE 1
REFERENCE MATERIAL REQUIREMENTS

Test items to support all aspects of the requalification examination must be provided to the NRC 60 days before the examination date. These may include the following:

(1)   Existing learning objectives, student handouts, and lesson plans (including training manuals, facility orientation manual, system descriptions, reactor theory, and thermodynamics principles).

  Training materials should include all substantive written material used for preparing applicants for initial RO and SRO licensing. The written material should include learning objectives, if available, and the details presented during lectures, rather than outlines. Training materials should be identified, bound, and indexed. Training materials that include the following should be provided:

  •   System descriptions including descriptions of all operationally relevant flowpaths, components, controls, and instrumentation. System training material should draw parallels to the actual procedures used for operating and to the applicable system.

  •   Complete and operationally useful descriptions of all safety system interactions, secondary interactions under emergency and abnormal conditions, including consequences of anticipated operator error, maintenance error, and equipment failure.

  •   Training material used to clarify and strengthen understanding of emergency operating procedures.

(2)   Complete procedure index (including temporary procedures).

(3)   All administrative procedures applicable to reactor operation or safety.

(4)   All integrated facility procedures, normal or general operating procedures, and procedures for experiments.

(5)   All emergency procedures, emergency instructions, and abnormal or special procedures.

(6)   Standing orders or procedures changed by reactor supervision and important orders or changes that are safety related and may supersede the regular procedures.

(7)   Applicable procedures (procedures that are run frequently).

(8)   Fuel-handling and core-loading procedures and initial core-loading procedures, when appropriate.

(9)   Any annunciator/alarm procedures, as applicable.

(10) Radiation protection manual, radiation control manual, or procedures.

(11) Emergency plan implementing procedures.

(12) Technical specifications and interpretations, if available.

(13)   System operating procedures, including experiments.

(14)   Piping and instrumentation diagrams, electrical single-line diagrams, or flow diagrams, as applicable.

(15)   Technical data book and/or facility curve information as used by operators and precautions, limitations, and set points for the facility.

(16)   Questions and answers specific to the facility training program, which may be used in the written or operating examinations (voluntary by facility licensee).

(17)   Additional material as requested by the examiners to develop examinations that meet the requirements of the research and test reactor examiner standards and regulations.

The above reference material should be the approved final issues and so marked. If a facility has not finalized some of the material, the NRC CE should verify with the facility that the most complete, up-to-date material is available and that agreement has been reached with the licensee on limiting changes before the administration of the examination.

ENCLOSURE 2
ADMINISTRATION OF REQUALIFICATION EXAMINATIONS

(1)   The CE will review the reference material.

(2)   The facility shall provide a single room for the written examination. The location of this room and supporting rest room facilities shall be chosen to minimize contact with other facility personnel for the duration of the examination.

(3)   The CE will ascertain that the facility provides the minimum spacing required to ensure examination integrity. Minimum spacing consists of one examinee per table and a 3-foot space between tables. No wall charts, models, and/or other training materials shall be present in the examination room.

(4)   Each examinee will receive copies of reference material for the written examination. The CE will review the reference material, which will consist of technical specifications, operating/abnormal procedures, administrative procedures, and emergency plans as available to the facility operators.

(5)   An attempt will be made to distinguish between RO and SRO knowledge and abilities to the extent that the facility training materials support such a distinction.

(6)   Prudent scheduling of the activities for examination week is important to help alleviate undue stress on the operators. Your training staff and the CE should work very closely in formulating a schedule that does not result in excessive delays of individuals being administered their examination.

ENCLOSURE 3
NRC RULES AND GUIDANCE FOR EXAMINEES

(1)   Print your name in the blank provided on the cover sheet of the examination.

(2)   Fill in the date on the cover sheet of the examination, if necessary.

(3)   Answer each question on the answer sheets.

(4)   The point value for each question is indicated in parentheses after the question.

(5)   If the intent of parts of the examination are not clear, ask questions of the proctor only.

(6)   You must sign the statement on the cover sheet that indicates the work on the examination is your own and that you have not received or given any assistance in completing the examination. This must be signed **after** the examination has been completed.

(7)   Rest room trips are to be limited, and only one examinee at a time may leave. You must avoid all contact with anyone outside the examination room to avoid even the appearance of examination compromise.

(8)   Cheating on the examination will result in a revocation of your license and could result in more severe penalties.

(9)   Each section of the examination is designed to take approximately 60 minutes to complete. You will have 3 hours to complete the examination.

(10)   When you are finished and have turned in your completed examination, leave the examination area.

ENCLOSURE 4
SECURITY AGREEMENTS

PREEXAMINATION SECURITY AGREEMENT

I _____ agree that I will not knowingly divulge any information
　　　　[Print Name]

concerning the requalification examination for _____ to any
　　　　　　　　　　　　　　　　　　　　　　　　　　[Print Facility Name]

unauthorized persons.

_____ / _____

Signature　　　　　　　　　　　　　Date

POSTEXAMINATION SECURITY AGREEMENT

I _____ did not, to the best of my knowledge, divulge any information
[Print Name]

concerning the examination administered on _____ to any unauthorized
　　　　　　　　　　　　　　　　　　　　　　　[Print Date]

persons.

_____ / _____

Signature　　　　　　　　　　　　　Date

ATTACHMENT 2
BRIEFING CHECKLIST—OPERATING TEST TASKS

(1)   The NRC examiner is a visitor; the facility is responsible for providing an escort to ensure compliance with safety, security, and radiation protection procedures.

(2)   Do not operate facility equipment unless **specifically allowed** by procedure, standing order, management direction, routine operating procedures, or other administrative allowances. If equipment operation is specifically prohibited, nothing I or the NRC examiner says or asks will be intended to violate that principle.

(3)   Do not hesitate to ask for clarification during the walk-through. You may request that the examiner or I reword or clarify the question.

(4)   We will be taking notes throughout the test to document your performance. Frequently, we will stop questioning for this purpose. The amount of note-taking does not reflect your level of performance. The examiner must document satisfactory as well as less than satisfactory performance.

(5)   The operating test is considered "open reference." The reference material in the facility/control room that is normally available to operators is available during the test. This includes calibration curves, previous log entries, piping and instrument diagrams, calculation sheets, and procedures. However, operators are responsible for knowing from memory the immediate actions of emergency and other procedures as appropriate to the facility.

(6)   The operating test has been planned to last approximately 90 minutes. However, there is no specific time limit. The examiner will take whatever time is necessary to cover the areas selected, in the depth and scope required. The test will evaluate a minimum of five tasks.

(7)   The examiner will explain the tasks to be completed and which steps to simulate or discuss and will provide initial conditions. The operator is to proceed with completing the task as if directed by facility procedures and/or shift supervision. During the task, the examiner will supply the necessary facility conditions and/or parameters needed to simulate the task. The operator should explain each step of the task to the examiner before doing it.

(8)   The examiner is not allowed to reveal the results of the examination at its conclusion.

(9)   The NRC examiner may ask clarifying questions of the operator at the end of each task. To the extent possible and reasonable, these questions will be asked via the facility evaluator.

(10)   The NRC examiner will indicate to the operator that no aspects of the examination should be discussed with anyone until the conclusion of the examination.

(11)   You may request a break at any time during the operating test.

(10)   During the examination you will be evaluated for your actions as if you were the actual watchstander. Please operate the reactor as if you were licensed, with the exception that you should announce your actions, then pause momentarily to give the operator of record time to correct you or stop you, if necessary, before you actually perform the action. In addition, the examiner will be observing that you meet all conditions of your license, (e.g., wearing corrective lenses to perform licensed duties).

FORM ES-601N-1
REQUALIFICATION EXAMINATION OPERATING TEST RECORD

Name:_____   License Number_____

	Critical Steps Number/Correct	Oral Questions Number/Correct
Control Console Reactivity Procedure No.: Title:	_____/_____	_____/_____
Response to Abnormal Event Procedure No.: Title:	_____/_____	_____/_____
Response to Emergency Procedure No.: Title:	_____/_____	_____/_____
In-Facility Evolution Procedure No.: Title:	_____/_____	_____/_____
Any Task Procedure No.: Title:	_____/_____	_____/_____
Totals	_____/_____	_____/_____
Percentage	_____	_____

Overall Evaluation: _____ Satisfactory   ≥80%   ≥70%

_____ Unsatisfactory

Facility Examiner _____ Date _____

NRC Chief Examiner _____ Date _____

FORM ES-601N-2
EXAMINATION COVER SHEET

U.S. NUCLEAR REGULATORY COMMISSION
RESEARCH AND TEST REACTOR REQUALIFICATION EXAMINATION

FACILITY:

REACTOR TYPE:

DATE ADMINISTERED:

CANDIDATE:_____

INSTRUCTIONS TO CANDIDATE:

Answers are to be written on the answer sheets provided. Attach all answer sheets to the examination. Points for each question are indicated in parentheses after each question. A score of 70 percent overall is required to pass the examination.

Examinations will be collected 3 hours after the examination starts.

CATEGORY VALUE	% OF TOTAL	CANDIDATE'S SCORE	% OF CATEGORY VALUE	CATEGORY	
20.00	33.33	_____	_____	A.	REACTOR THEORY, THERMODYNAMICS, AND FACILITY OPERATING CHARACTERISTICS
20.00	33.33	_____	_____	B.	NORMAL AND EMERGENCY PROCEDURES AND RADIOLOGICAL CONTROLS
20.00	33.33	_____	_____	C.	FACILITY AND RADIATION MONITORING SYSTEMS
60.00	100.00	_____	_____ %	FINAL GRADE	

All work done on this examination is my own. I have neither given nor received aid.

_____
Candidate's Signature

Form ES-601N-3
Research and Test Reactor Requalification Results Summary Sheet
Personally Identifiable Information
— Withhold under 10 CFR 2.390

RESEARCH AND TEST REACTOR REQUALIFICATION RESULTS SUMMARY SHEET

FACILITY:

EXAM DATE:

EXAMINERS:

OVERALL RESULTS	TOTAL	PASSED	FAILED
	#	# / %	# / %
REACTOR OPERATOR		/	/
SENIOR OPERATOR		/	/
TOTAL		/	/

NAME	DOCKET 55-	GRADER	TASK _OF_	TASK % OVERALL	TASK QUESTIONS _OF_	TASK % QUESTIONS OVERALL	WRITTEN % A	B	C	TOTAL	RESULTS/INITIALS WRITTEN	OPERATING
		NRC	5 OF 5		OF		100	100	100	100	/	/
		FAC	OF		OF						/	/
		NRC	OF		OF						/	/
		FAC	OF		OF						/	/
		NRC	OF		OF						/	/
		FAC	OF		OF						/	/
		NRC	OF		OF						/	/
		FAC	OF		OF						/	/
		NRC	OF		OF						/	/
		FAC	OF		OF						/	/
		NRC	OF		OF						/	/
		FAC	OF		OF						/	/

FORM ES-601N-4
INDIVIDUAL REQUALIFICATION EXAMINATION REPORT

INDIVIDUAL REQUALIFICATION EXAMINATION REPORT

FACILITY:	EXAMINEE'S NAME:	
DOCKET NO. 55-	LICENSE NO.:	EXPIRATION DATE:
EXAM TYPE: RO/SRO	RETAKE EXAM: 1ST/2ND/NO	DATE OF LAST EXAM (RETAKE):

EXAM SUMMARY		
WRITTEN EXAM RESULTS		
DATE OF EXAM:	NRC EXAMINER (PRINT):	FACILITY EXAMINER (PRINT):
	NRC GRADING	FACILITY GRADING
SECTION A (POINTS)	OF %	OF %
SECTION B (POINTS)	OF %	OF %
SECTION C (POINTS)	OF %	OF %
OVERALL SCORE	%	%

OPERATING EXAM RESULTS		
DATE OF EXAM:	NRC EXAMINER (PRINT):	FACILITY EXAMINER (PRINT):
NO. CORRECT TASKS	OF %	OF %
NO. QUES. CORRECT	OF %	OF %

NRC EXAMINER RECOMMENDATIONS		
CATEGORY	RESULTS	SIGNATURE
WRITTEN	PASS/FAIL	
OPERATING	PASS/FAIL	

CHIEF EXAMINER REVIEW		
PASS/FAIL	SIGNATURE:	DATE:

EXAMINER STANDARD 602N
REQUALIFICATION FAILURES, REVIEWS, AND HEARINGS

A. Purpose

The NRC staff should use these procedures to process requests for administrative reviews and hearings by 10 CFR Part 55 licensees concerning failures of NRC-conducted requalification examinations and denials of applications for license renewal.

B. Background

The renewal license application differs in some respects from the initial license. To address these differences, the staff has developed these procedures for requests for administrative reviews and hearings for the denial of license renewal.

C. Results of NRC-Conducted Requalification Examinations

1. Passing an NRC-Conducted Requalification Examination

The BC/PD will inform all operators who pass all portions of the requalification examination in a letter in the format of Attachment 1.

2. Failure of Written Examination Section(s)

If an operator failed one or two sections but received an overall grade of 70 percent or better, the facility will determine the appropriate corrective action and retest the operator on the failed section(s). No further NRC involvement is necessary.

3. Failing an NRC-Conducted Requalification Examination

a.   If an operator fails any part of an NRC-conducted requalification examination, the facility licensee must remove that operator from licensed duty. The facility licensee must take corrective action consistent with the provisions of its requalification program before returning the operator to licensed duty.

b.   The BC/PD will inform the operator of his or her failure in a letter in the format of Attachment 2 or Attachment 3 of this standard as appropriate. The operator then has 20 days to request an informal review of the failed portion of the examination. The operator may submit a request for reconsideration to the appropriate Division Director, U.S. Nuclear Regulatory Commission, Washington, DC 20555. Note 5 in Section E provides additional information on the requirements for supporting documentation.

c.   The NRC will not renew the license of any operator who failed to pass an NRC-conducted requalification examination, without some level of NRC involvement in the retesting process. NRC involvement may include conducting a retest according to this NUREG, inspecting the facility as it retests the operator, or simply reviewing an examination prepared by the facility. The BC/PD will determine the appropriate level of involvement depending on the quality of the facility's program. As long as the operator submits a timely renewal application, the term of the operator's license will continue until the renewal requirements are satisfied or until the operator has failed three NRC-conducted examinations as discussed in Section C.2.e.

d.   The NRC will normally conduct a second (first retake) examination approximately 6 months after issuing the first failure notification according to Section C.3.b of this standard and will concentrate on the areas in which the operator exhibited deficiencies.

e.   The NRC will conduct a third (second retake) examination approximately 6 months after issuing a second failure notification according to Section C.3.b of this standard. The third examination will be a **comprehensive** requalification examination.

      If an operator fails a third requalification examination, the NRC will review the operator's performance and the facility's training program. A third failure may be grounds for suspending or revoking the operator's license. If an operator has an application pending for license renewal at the time of a third requalification failure, that failure will provide the basis for denying the application. Notification of the operator will be handled on a case-by-case basis and coordinated through the BC/PD.

D. Overturning Requalification Examinations or Renewal Denials

If, upon conducting a hearing or an informal review, the staff reverses its decision regarding the failure of a requalification examination or application denial, the staff will take one or more of the following actions, as appropriate:

(1)   reinstate the license

(2)   allow the operator to renew the license pursuant to 10 CFR 55.57, if all other requirements are satisfied

(3)   allow the operator to perform licensed duties when he or she has successfully completed the facility's requalification program and the provisions of 10 CFR 55.53(e) or (f)

If, upon conducting a hearing, the staff reverses its decision regarding the failure of the requalification examination, the staff will inform the operator that he or she has passed the examination using a letter similar in format to the letter in Attachment 1.

If, upon conducting a hearing, the staff reverses its denial of an operator's renewal application, the operator will be eligible for license renewal pursuant to 10 CFR 55.57 if the licensee has satisfied all other requirements that were not at issue in the hearing.

E. Notes

(1)   Letters informing an operator of a proposed denial or examination failure must be signed at the division level and a copy of the correspondence will be distributed to the branch (program) level for tracking purposes.

(2)   Copies of correspondence sent to the operator because of this process will be provided to the facility licensee's representative who is authorized to sign the renewal application.

(3)   All examination failure or application denial correspondence sent to the operator should be sent by certified mail with return receipt requested.

(4)   Asking the facility licensee to reassess the need for an operator's license is inappropriate while conducting an informal review or hearing.

(5)   Requests for informal review by the NRC must (a) list the items for which additional review is being requested and (b) include documentation supporting the contentions made by the operator. The package containing the review request along with supporting documentation must be mailed or delivered to the division director's office within 20 days of the date of the failure or denial notification. Division staff should complete the review within 45 days of receiving the package. The staff will review requests using the guidance in ES-502N.

ATTACHMENTS/FORMS:

Attachment 1, Requalification Examination Pass Letter
Attachment 2, Requalification Examination Failure Letter
Attachment 3, Requalification Examination Second Failure Letter

ATTACHMENT 1
REQUALIFICATION EXAMINATION PASS LETTER

[Date]

Docket No. 55-[Number]

[Applicant/Facility Name]
[Street Address]
[City, State Zip Code]

Dear [Name]:

I am writing to inform you that you passed the requalification written examination and operating test conducted by the U.S. Nuclear Regulatory Commission (NRC) on [date]. Enclosed is a copy of your "Individual Requalification Examination Report" (Form ES-601N-5) summarizing the results of your examination. Your facility training department has a copy of the master answer key.

If you have a question, please contact [name] at [telephone number].

Sincerely,

[BC/PD or higher]

cc: [Facility-authorized representative who signs NRC Form 398]

ATTACHMENT 2
REQUALIFICATION EXAMINATION FAILURE LETTER

[Date]

Docket No. 55-[Number]

[Applicant/Facility Name]
[Street Address]
[City, State Zip Code]

Dear [Name]:

I am writing to inform you that you did not achieve an acceptable score on the requalification [written examination and/or operating test] conducted by the U.S. Nuclear Regulatory Commission (NRC) on [date]. Enclosed is a copy of the results indicating the area(s) in which you exhibited deficiencies. Your facility training department has a copy of the master answer key.

This failure places you in the same status as if you had failed a facility-conducted requalification examination. Therefore, you are subject to the requirements set forth in the NRC-approved requalification program for the facility for which you are licensed and must meet those requirements before resuming licensed duties. The NRC will conduct a second requalification examination in the areas in which you exhibited deficiencies.

If you believe an error was made in grading your examination, you may request within 20 days of the date of this letter that the NRC informally regrade the examination. Requests for informal regrade should be sent to the [appropriate division director], U.S. Nuclear Regulatory Commission, Washington, DC 20555. In the request, please state the items you wish to have reviewed and provide supporting documentation as applicable.

If you have any questions, please contact [name] at [telephone number].

        Sincerely,

        [BC/PD or higher]

cc: [Facility authorized representative who signs NRC Form 398]

CERTIFIED MAIL—RETURN RECEIPT REQUESTED

ATTACHMENT 3
REQUALIFICATION EXAMINATION SECOND FAILURE LETTER

[Date]

Docket No. 55-[Number]

[Applicant/Facility Name]
[Street Address]
[City, State Zip Code]

Dear [Name]:

Based on the grading of the NRC-conducted requalification [written examination and/or operating test] taken on [date], you did not achieve an acceptable score. Enclosed is a copy of the results indicating the area(s) in which you exhibited deficiencies. Your facility training department has a copy of the master answer key.

This failure places you in the same status as if you had failed a facility-conducted requalification examination. Therefore, you are subject to the requirements set forth in the NRC-approved requalification program for the facility for which you are licensed and you must meet those requirements before resuming licensed duties. The NRC will conduct a third requalification examination that will be comprehensive in scope.

If you believe an error was made in grading your examination, you may request within 20 days of the date of this letter that the NRC informally regrade the examination. Requests for informal regrade should be sent to the [appropriate division director], U.S. Nuclear Regulatory Commission, Washington, DC 20555. In the request, please state the items you wish to have reviewed and provide supporting documentation as applicable.

If you have any questions, please contact [name] at [telephone number].

    Sincerely,

    [BC/PD or higher]

cc: [Facility authorized representative who signs NRC Form 398]

CERTIFIED MAIL—RETURN RECEIPT REQUESTED

EXAMINER STANDARD 701N
SRO LIMITED TO FUEL HANDLING—EXAMINATIONS

A. Purpose

This standard gives guidelines and instructions for preadministration activities in support of LSRO examinations at research and test reactors that have been permanently shut down. This standard supplements guidance contained in ES-201N, ES-202N, ES-203N, and ES-204N for administering initial examinations.

B. Examination Coordination

The CE will schedule a specific examination date with the facility staff. Normally, examiners will administer all examinations in 1 week, giving the written examination before administering the operating tests.

At least 3 months before the examination week, the CE will reconfirm the date of the examination and the number of candidates to be examined. The CE should use ES-201N, Attachment 1 ("Sample Corporate Notification Letter"), as a guide and discuss the following:

(1)   the need to have the reference materials identified in Attachment 1 at least 60 days before the scheduled examination date

(2)   the guidelines for review and administration of the written examinations (Enclosures 1, 2, 3, and 4 in Attachment 1)

(3)   the requirements (10 CFR 55.31) and guidelines (Attachment 1) for submitting the license applications

The CE will normally issue a letter confirming these arrangements about 90 days before the examination begins. This letter should be addressed to the person at the highest level of management responsible for facility operations. Attachment 1 is an example of such a letter. The CE may modify the wording as necessary to reflect the situation. The assignment of examiners should follow the guidelines of ES-201N, Section D.

If the examination must be rescheduled on short notice, the CE should reschedule to reduce the effect on other examinations scheduled. An examiner who fails an applicant on an operating test may not administer that candidate's retake operating test.

C. Eligibility

The duties and responsibilities associated with maintaining the facility after permanent shutdown are much less than those associated with an operating reactor. Therefore, the eligibility requirements for an LSRO are less than those for an SRO at an operating reactor.

(1)   Education

The educational requirements are the same as described in ES-202N, Section E.1(c).

---

(2)   Training

The applicant must satisfactorily complete a training program covering fuel-handling operations. The program should include instruction in the following areas:

a.   health physics fundamentals and the principles of reactor theory and thermodynamics

b.   design features of fuel-handling and storage activities and conditions, including facility systems and equipment associated with fuel-handling operations, pertinent instrumentation, and control systems

c.   use of systems to control or mitigate an accident in which the fuel is damaged

d.   administrative, operational, surveillance, emergency, and radiation control, security, and safety procedures concerning fuel handling and storage

In addition, the applicant should have actively participated in at least one fuel-handling evolution (this can be a dummy fuel movement) at the site for which the license is sought or at a similar facility. The applicant should also complete a minimum of 10 hours of on-the-job training in fuel-moving activities, including manipulation of the refueling equipment.

Persons who have either RO or SRO licenses at the time the reactor is shut down will be renewed as LSROs.

D. Reviewing LSRO Initial License Applications

The regulatory requirements associated with the license process are as stated in 10 CFR Part 55, Subpart D The CE should refer to them as necessary when reviewing license applications.

(1)   Each applicant must submit an NRC Form 398 and an NRC Form 396. (Computer-generated duplicates are acceptable.) An application is not complete until both forms are filled out, signed by the appropriate personnel, and signed by the NRC.

Detailed instructions for completing NRC Form 398 accompany the form.

(2)   The facility licensee's senior management representative must certify that the applicant has completed the training required for the desired license level by placing a check in Item 19.b and signing the application. He or she should submit the completed form to the applicable BC/PD, at least 14 days before the examination date.

(3)   The CE will review the applications against the eligibility requirements described in this standard, process the medical certifications and any waiver requests (refer to ES-203N), and request any additional information that might be necessary.

(4)   If the CE decides that an application does not meet the requirements of 10 CFR 55.31, as modified by this standard, he or she will note the deficiencies and ask the facility licensee to supply additional information. If the CE decides that the applicant still does not meet the eligibility requirements, the BC/PD will notify the applicant in writing that the application is being denied and inform him or her of the deficiencies on which the denial is based (Attachment 2).

In addition, the CE will check the *Does Not Meet Requirements* block at the bottom of the application and sign and date the form. An applicant will not be permitted to take a license examination until he or she meets the eligibility requirements.

If the applicant does not accept the proposed denial, he or she may request that the applicable division director review the application denial or may request a hearing according to 10 CFR 2.103(b)(2). Further action will be taken according to ES-502N.

E. Medical Requirements

See ES-202N, Section D.

F. NRC Form 398

Each applicant must submit an NRC Form 398. The form must be completely filled out per the instructions and signed by the appropriate personnel. Those sections or items that are not applicable to operators at research and test reactors that have ceased operations will be marked "NA."

G. Maintaining Medical Standards for Licensees

See ES-202N, Section G.

H. Routine Waivers

See ES-203N, Section D.

I. Routine Exemptions

Because the facility is permanently shut down, it is impossible for the applicant to meet the requirements of 10 CFR 55.31(a)(5). All applicants for an LSRO license at a permanently shutdown facility should request an exemption to this requirement. This exemption **will** be granted for all applicants at permanently shutdown reactors.

ATTACHMENTS/FORMS:

Refer to Sample Corporate Notification Letter, Attachment 1 to ES-201N

EXAMINER STANDARD 702N
SRO LIMITED TO FUEL HANDLING—EXAMINATION PREPARATION

A. Purpose

This standard specifies guidelines and instructions for preparing LSRO examinations. This standard supplements guidance contained in ES-301N, ES-303N, ES-401N, ES-402N, and ES-403N for administering initial examinations.

B. Scope

The NRC's LSRO examination comprises a written examination and an operating test. This standard includes the requirements for the preparation of these examinations.

C. General Guidelines

(1)   The examination will consist of both a written examination and an operating test. Applicants must pass both to obtain a license. The written examination and operating test will be administered as described in ES-703N. If the facility licensee has a refueling simulator, or dummy fuel elements, the examiner should use them to the maximum extent possible.

(2)   Examinations will be documented using the same approval chain as initial examinations and use the cover sheet (Form ES-702N-1) included in this standard.

(3)   If an applicant applies for a license on multiple units at a site, the examiner should ensure comprehensive coverage for all of the applicable units.

(4)   Additional unit licenses may be issued if the applicant passes a plant difference examination.

(5)   LSRO licenses will be renewed on the same basis as any other operator's license.

D. Written Examination Instructions

1. Content and Preparation

LSRO written examinations should consist of questions totaling 20±3 points and should be constructed so that a competent applicant can complete the examination in 40 minutes. Applicants will be allowed a maximum of 1 hour to complete and review the examination.

The written examination will contain questions covering the 21 items specified in 10 CFR 55.41 and 10 CFR 55.43 as they relate to the shutdown reactor.

If the reference material supplied by the facility licensee does not enable the examiner to construct a balanced examination, the examiner should contact the facility licensee to obtain additional material.

2. Quality Assurance Review

The written examination should meet all the guidelines and requirements for test question construction, quality assurance, and facility reviews specified in ES-401N and NUREG BR-0122, "Examiners' Handbook for Developing Operator Licensing Written Examinations," Revision 5, issued in 1990.

3. Administration

The written examination should be administered according to ES-402N, except that the time limit to complete the examination is 1 hour.

E. Operating Test Instructions

The operating test should be performance based. The facility licensee should be encouraged to permit the actual use of equipment to handle dummy fuel elements, assemblies, or modules during the operating test whenever feasible. This may require careful coordination with the facility licensee.

The examiner will assess the applicant's ability to execute normal, abnormal, and emergency procedures associated with fuel handling. Each applicant will be required to simulate or perform tasks related to fuel handling and to answer questions associated with refueling equipment and associated systems. The operating test will also determine whether the applicant can supervise the operation of equipment and systems to conduct fuel-handling operations safely.

1. Administrative Topics (Category A)

Category A covers topics associated with the administrative control of the facility divided into three groups. The examiner must cover the minimum number of subjects from each group shown in the heading on page 2 of Form ES-702N-1. Selection of subjects is at the discretion of the examiner. To evaluate varied subjects with different applicants, the examiner should be familiar with all of the administrative topics.

In developing Category A questions, the examiner may use previously developed questions from either an NRC or facility examination bank or may develop new open-reference questions. In any case, the questions asked should be pre-scripted to the maximum extent possible. Any questions asked to follow up on perceived applicant weaknesses must be documented for grading and review.

The examiner should use the following descriptions as guidelines for developing or selecting questions to confirm minimal competency within each subcategory:

a.   Subcategory A.1

Questions in Subcategory A.1 will evaluate the applicant's knowledge of the daily administrative operation of the facility. The questions may be integrated into other discussions as they apply throughout the test.

b.   Subcategory A.2

These subjects deal with radiation protection, including the approval of release permits and awareness of the requirements associated with those releases and their potential effect on the health and safety of the public. A task with followup questions is appropriate for performing an evaluation of this subcategory.

c.   Subcategory A.3

Applicants must display knowledge based on their responsibility to direct and manage the EPIPs during the initial phases of an emergency. Applicants should be familiar with event classification procedures, communication requirements and methods, and have a detailed understanding of the EPIPs overall.

Security may be evaluated by observing the applicant's behavior during the examination. Questioning an applicant about applicable aspects of the facility's security plan is appropriate, however.

2. Facility Walkthrough (Category B)

This category tests the applicant's knowledge of system design and operation. The examiner will evaluate the applicant's ability to perform tasks and to answer questions about specific systems.

Attachment 1, "Systems for Operating Tests," lists systems typically found at research and test reactors. The examiner should select systems from this list as applicable to the specific facility. To enhance test integrity, the examiner should vary coverage of systems and subjects across test administrations. The examiner will evaluate applicants on at least five systems from at least three categories listed in Attachment 1. Note that Attachment 1 may not be all inclusive.

Questions asked to clarify the performance of an applicant must be documented for postexamination evaluation.

The examiner may select tasks and pre-scripted questions from existing facility examination banks. The examiner may choose to annotate procedures with clarifying comments on how to execute particular steps, as well as identifying critical steps. To evaluate a subject area satisfactorily, the examiner will ask enough questions to determine the applicant's knowledge.

If the facility's examination bank is used, no more than 40 percent of the questions associated with a particular system may be used without some change.

ATTACHMENTS/FORMS:

Attachment 1, Systems for Operating Tests
Form ES-702N-1, Senior Reactor Operator Limited to Fuel Handling Examination Report

ATTACHMENT 1
SYSTEMS FOR OPERATING TESTS

MAJOR SYSTEMS
Makeup Water/Purification System
Reactor Pool System
Fuel Storage Pool System
Neutron Source

WASTE-HANDLING FACILITIES
Liquid Waste System
Solid Waste System
Gaseous Waste System

INSTRUMENTATION SYSTEMS
Area Radiation Monitoring
Gaseous Radiation Monitoring
Liquid Effluent Radiation Monitoring

AUXILIARY SYSTEMS
Normal AC Supply
Batteries
Service Air
Reactor Building Air Recirculation
Fuel Handling
Containment/Reactor Building Isolation

NRC FORM ES-702N-1   U.S. NUCLEAR REGULATORY COMMISSION

SENIOR REACTOR OPERATOR LIMITED TO FUEL HANDLING
EXAMINATION REPORT

CANDIDATE'S NAME	DOCKET NUMBER 55-7XXXX	FACILITY

EXAMINATION TYPE

SENIOR REACTOR OPERATOR LIMITED TO FUEL HANDLING

INITIAL EXAMINATION	.	RETAKE EXAMINATION

WRITTEN EXAMINATION SUMMARY

	GRADE
Written by: (Print Name)	
Date Administered	
Graded by: (Print Name)	%

OPERATING TEST SUMMARY

Administered by:	Date Administered:
A. Administrative Topics SAT/UNSAT	B. Facility Walk-Through/Integrated Plant Operations SAT/UNSAT

EXAMINER RECOMMENDATIONS

	PASS	FAIL	WAIVE		
Written Examination				GRADER	DATE:
Operating Test				ADMINISTRATOR	DATE:
Final Recommendation				PEER REVIEWER	DATE:

LICENSE RECOMMENDATION

		SIGNATURE CHIEF EXAMINER:	DATE:
	ISSUE LICENSE		
	DENY LICENSE		

A. ADMINISTRATIVE TOPICS			
A.1 (MINIMUM 3)	EVALUATION	A.2 (MINIMUM 2)	EVALUATION
A. CONDUCT OF OPERATIONS	SAT UNSAT N/E	A. RADIATION SOURCES AND HAZARDS AND DETECTION	SAT UNSAT N/E
B. MODIFICATIONS (SYSTEMS AND PROCEDURES)	SAT UNSAT N/E	B. RADIATION EXPOSURE LIMITS AND CONTAMINATION CONTROL	SAT UNSAT N/E
C. SURVEILLANCE TESTING	SAT UNSAT N/E	C. RADIATION WORK PERMITS, RADIATION RELEASE CONTROL (PERMITS, RATES, LIMITS)	SAT UNSAT N/E
D. MAINTENANCE PROCEDURES	SAT UNSAT N/E	A.3 (ALL MANDATORY)	EVALUATION
E. SHORT-TERM INFORMATION (NIGHT/STANDING ORDERS, ETC.)	SAT UNSAT N/E	A. EMERGENCY PLAN (UNDERSTANDING/ FAMILIARITY)	SAT UNSAT N/E
		B. SECURITY (AWARENESS/ FAMILIARITY)	SAT UNSAT N/E

B. FACILITY WALKTHROUGH (MINIMUM 5)	TYPE	SUBJECT A:* EQUIPMENT/COMPONENTS	SUBJECT B:* INSTRUMENTATION/PROTECTION/INTERLOCKS	SUBJECT C:* PROCEDURAL KNOWLEDGE/USE	SUBJECT D:* ADMINISTRATIVE REQUIREMENTS	SYSTEM GRADE	COMMENT PAGE NO.
SYSTEM/TASK TITLE							
1							
2							
3							
4							
5							
6							
7							
8							
9							
10							

OPERATING TEST COMMENTS

Alpha/Numeric Subject Index	Comment

EXAMINER STANDARD 703N
SRO LIMITED TO FUEL HANDLING—EXAMINATION ADMINISTRATION

A. Purpose

This standard specifies guidelines and instructions for administering LSRO examinations. This standard supplements guidance contained in ES-302N and ES-402N.

B. Examination Withdrawals

Occasionally, an applicant will withdraw from the examination just before its start. When this happens, the examiner will request a letter from the facility withdrawing the application of the individual(s).

In rare instances, an applicant will withdraw after the examination has begun. The examiner will inform the applicant that this is an examination failure, and he or she must reapply following the rules of 10 CFR 55.35.

C. Written Examination Facilities

(1)   The facility licensee is responsible for providing facilities suitable for administering the written examination. The room and associated rest room(s) should allow the NRC examiners to maintain examination security. Enclosure 2 in Attachment 1 to ES-701N, "Administration of Written Examinations," summarizes the NRC's policies regarding written examination facilities.

(2)   The CE will evaluate the facilities to assure that the applicants do not have access to any reference material not approved by the CE. The CE will not begin the examination until he or she is satisfied with the arrangements.

D. Proctoring the Written Examination

(1)   The CE will ensure that the examination is proctored at all times. Before distribution of examinations, the CE will ensure that all proctors clearly understand their responsibilities.

If the CE determines that he or she needs assistance with the administration or the oversight of the administration of the written examination, the CE should request assistance from another examiner or other responsible NRC employee.

The proctor will not engage in any activities that may divert his or her attention from the applicants and possibly cause the examination to be compromised.

(2)   An examiner should be available to clarify questions for the applicants during the examination. Proctors should be careful not to give away answers when clarifying questions. If a proctor has any doubt about how to respond to an applicant's question, he or she should consult with the CE before explaining the test item to the applicant. Proctors will document all questions regarding specific written examination test items for future reference in resolving facility comments and grading conflicts.

When responding to questions, the proctor should be alert for indications that an applicant is unfamiliar with the terminology used in the examination. The proctor will ask the CE to determine the correct terminology and announce it to all the applicants taking the examination.

All question changes or clarifications will be called to the attention of all the applicants. Changes made to questions during the examination should be made on the master copy and on the copy provided to the facility staff.

E. Written Examination Administration Procedure

The CE will administer the written examinations as follows:

(1)   Verify each applicant's identity and examination level against the examination assignment sheet (see Form ES-701N-1). Any errors or absences will be resolved with the facility staff, and the assignment sheet will be updated as required.

        The CE will request the facility licensee formally to withdraw the application of any individual not taking the examination by submitting a letter to the BC/PD.

(2)   Remind the applicants that they may use calculators to complete the examination. Only reference material approved by the examiner is allowed in the examination area. The examiner will define the examination area.

(3)   Pass out the examinations, answer sheets, and handouts and instruct the applicants not to review the examination until told to do so.

(4)   Brief the applicants on the rules and guidelines that will be in effect during the written examination using Enclosure 3 in Attachment 1 to ES-701N, "Policies and Guidelines for Taking NRC Written Examinations." Inform the applicants that they may refer to the instructions directly beneath the examination cover sheet. Read the first two policies **verbatim**.

(5)   Ask the applicants to verify the completeness of their examination by checking each page.

(6)   After answering any questions that the applicants may have regarding examination policies, start the examination and record the time.

(7)   Periodically advise the applicants of the time remaining.

(8)   Ensure that each applicant signs the cover sheet when turning in his or her examination. The proctor will ensure that the examination package is complete (includes the cover sheet and answer sheets).

(9)   Remind the applicants to leave the examination area.

(10)   When all of the written examinations are complete, the CE may conduct an examination review with the facility staff as described in Section F below.

The CE will complete Form ES-703N-1, "Examination Administration Quality Assurance Checkoff Sheet," and include it in the master examination package.

F. Facility Staff Review of the Written Examination

Immediately after the last applicant completes the examination, the CE will update the master copy of the examination and answer key with all the changes made to questions while the examinations were being administered.

The CE may provide a copy of the master examination and answer key to the facility licensee staff and answer any questions they may have regarding the NRC's examination review and comment process. ES-201N, Enclosure 4, provides detailed guidelines and instructions for this review process.

G. Personnel Present at the Operating Test

The number of persons present during administration of an operating test should be limited to ensure the integrity of the examination and to reduce distractions to the applicant.

Only the CE may grant permission for someone to witness an operating test. Under **no** circumstances may another applicant witness an operating test. Operating tests are not to be used as training vehicles for future applicants. Videotaping of initial examinations is not allowed.

Examiners may witness an operating test as part of their training or may audit an examiner administering the operating test. Other observers, such as regional personnel, researchers, or NRC supervisors, may be allowed to observe operating examinations if (1) the CE has approved the request to observe the test and (2) the applicant does not object to the observer's presence.

H. Operating Test Administration Procedures

1. General

The examiner will brief each candidate according to Attachment 1 to this standard before beginning the operating test.

2. Administrative Topics

a.   Subcategory A.1

These questions are intended to supplement not duplicate administrative system requirements in Category B (e.g., valve lineups, control room data system administration and use).

b.   Subcategory A.2

These subjects are best covered during the conduct of tasks or questioning associated with Category B.

d.   Subcategory A.3

The examiner may best evaluate the applicant's knowledge of the emergency plan by conducting a Category B task requiring its use.

3. Operating Test

The examiner should encourage the applicant to draw diagrams, flowpaths, and other visual representations. Likewise, the examiner should encourage the applicant to use facility forms, schedules, procedures, and other similar materials as appropriate.

The examiner should retain any supporting material to provide additional documentation in support of a pass or fail determination.

The examiner can improve the efficiency of the operating test by integrating the discussions required to complete the various categories.

The examiner must take sufficient notes during the operating test to document all applicant deficiencies thoroughly. The examiner must cross-reference every comment to a specific task or subject area question.

ATTACHMENTS/FORMS:

Attachment 1, Operating Test Briefing Checklist

ATTACHMENT 1
OPERATING TEST BRIEFING CHECKLIST

Part A—Applicable to All Operating Tests

(1)   You will be tested at the level of responsibility of the senior licensed shift position (i.e., shift supervisor, senior shift supervisor, or whatever the position may be titled).

(2)   I am a visitor. Escort responsibilities for ensuring compliance with safety, security, and radiation protection procedures rest with you (the applicant).

(3)   Do not operate facility equipment without appropriate permission.

(4)   Do not hesitate to request clarification of a question during the operating test.

(5)   Frequently, I will stop to update my notes to document your performance. The amount of note-taking is not dependent on your level of performance.

(6)   Operating tests are considered "open reference." Any reference material in the facility normally available to operators is also available to you, including calibration curves, previous log entries, piping and instrumentation diagrams, calculation sheets, and procedures. However, you are responsible for knowing systematic automatic actions, set points and interlocks, operating characteristics, and the immediate actions of emergency and other procedures, as appropriate to the facility.

(7)   There is no specific time limit for the operating test. The exam will take whatever time is necessary to cover the areas selected, in the depth and scope required.

(8)   I am not allowed to reveal the results of the operating test at its conclusion.

(9)   Do not hesitate to request a break during the operating test.

Part B—Just before Fuel Movement

The examiner should brief the candidate and the licensed operator on the following points:

(1)   I will not intentionally ask you (the applicant) to do an act that violates facility regulations or procedures or which places the facility in a hazardous condition. If a requested act meets one of these conditions, then you (the applicant or operator) should immediately inform me. If my intent was to discover whether you (the applicant) would perform such an act, I will phrase the question in some manner other than a request that the act be performed.

(2)   My presence does not alter the normal chain of command during the examination. You (the candidate) should make all reports and obtain all permissions that would normally be required. All directions to the applicant will come from the responsible supervisor and follow the facility administrative procedures. The examiner will only question and make requests of the supervisor.

(3)   I have not altered the set points or calibrations of any instrument nor have I manipulated any controls.

(4)   It is your (the licensed operator's) responsibility to step in and take control any time there is an unsafe condition. However, you may not provide any coaching or cuing to the applicant.

EXAMINER STANDARD 704
SRO LIMITED TO FUEL HANDLING—POSTEXAMINATION ACTIVITIES

A. Purpose

This standard specifies guidelines and instructions for grading and documenting LSRO examinations. This standard supplements guidance contained in ES-303N and ES-403N for grading initial examinations.

B. Written Examination—Resolving Facility Comments

The CE should resolve facility comments per the instructions contained in ES-403N, Section B, paragraphs 1 through 4.

C. Written Examination—Grading

The CE will grade the examination following steps 1 through 5 of ES-403N, Section C. After grading the last applicant, the examiner will review the grading for all applicants in detail.

The grader will fill in the *Written Examination Summary* section of Form ES-704-1 and check the *Pass* or *Fail* or *Waive* block and sign and date the grader line in the *Examiner Recommendations* section. The grader will then forward the examination package to the person who administered the operating tests.

D. Operating Test—General Evaluation Guidelines

The operating test grader will score the test according to the guidelines in ES-303N, Section B.

E. Operating Test—Specific Instructions for Completing Form ES-902-1

The operating test grader will score the operating test and fill out the form following the guidelines of ES-303N, Section C, parts 1, 2, and 4 .

ATTACHMENTS/FORMS:

Form ES-704-1, Research and Test Reactor LSRO Examination Results Summary Sheet

PERSONALLY IDENTIFIABLE INFORMATION—
WITHHOLD UNDER 10 CFR 2.390

FORM ES-704-2
RESEARCH AND TEST REACTOR
LSRO EXAMINATION RESULTS SUMMARY SHEET

Facility:			Plant Status: PERMANENTLY SHUTDOWN		
Written Exam Date:			Operating Exam Dates:		
Examiners:					

RESULTS	TOTAL No.	No. Passed	% Passed	No. Failed	% Failed

Name	Docket 55-	Type[1]	Written Grade	Results[2]/Initials[3] Written	Operating

NOTES:
[1] 1 = LSRO initial, 2 = LSRO retake
[2] P = Passed, F = Failed, W = Waived, N/A = Not Applicable
[3] Enter only the initials of the examiner who actually wrote or administered the examination.

NRC FORM 335 (9-2004) NRCMD 3.7	U.S. NUCLEAR REGULATORY COMMISSION	1. REPORT NUMBER (Assigned by NRC, Add Vol., Supp., Rev., and Addendum Numbers, if any.)
	BIBLIOGRAPHIC DATA SHEET *(See instructions on the reverse)*	NUREG-1478 Rev. 2

2. TITLE AND SUBTITLE

Operator Licensing Examiner Standards for Research and Test Reactors

3. DATE REPORT PUBLISHED	
MONTH	YEAR
June	2007

4. FIN OR GRANT NUMBER

5. AUTHOR(S)

P. Doyle

6. TYPE OF REPORT

Staff Technical

7. PERIOD COVERED *(Inclusive Dates)*

8. PERFORMING ORGANIZATION - NAME AND ADDRESS *(If NRC, provide Division, Office or Region, U.S. Nuclear Regulatory Commission, and mailing address; if contractor, provide name and mailing address.)*

Divsion of Policy and Rulemaking
Office of Nuclear Reactor Regulation
U.S. Nuclear Regulatory Commission
Washington, DC 20555-0001

9. SPONSORING ORGANIZATION - NAME AND ADDRESS *(If NRC, type "Same as above"; if contractor, provide NRC Division, Office or Region, U.S. Nuclear Regulatory Commission, and mailing address.)*

Same as above

10. SUPPLEMENTARY NOTES

Revision 2 is a complete rewrite of NUREG-1478. Replace any old copies of NUREG-1478 with Revision 2. To be published June 2007

11. ABSTRACT *(200 words or less)*

NUREG-1478 "Operator Licensing Examiner Standards for Research and Test Reactors" establishes the policy, guidance and procedures for examining licensees and applicants for reactor operator and senior reactor operator licenses at non-power reactor facilities pursuant to Part 55 of Title 10 of the Code of Federal Regulations (10CFR55).

The NRC is issuing Revision 2 primarily to classify Rearch and Test Reactor (RTR) facilities into three new classes, Complex, Moderate and Simple. The revision presents new procedures for the generation, administration and grading of the written and operating tests at the three classes of facility types. In addition it clarifies the grading criteria for operating test portion of the examination. Finally a new series (700) covers the administration of Senior Reactor Operator licenses limited to fuel handling (LSRO).

12. KEY WORDS/DESCRIPTORS *(List words or phrases that will assist researchers in locating the report.)*

Operator Licensing
Examination
Research and Test Reactor

13. AVAILABILITY STATEMENT

unlimited

14. SECURITY CLASSIFICATION

(This Page)

unclassified

(This Report)

unclassified

15. NUMBER OF PAGES

16. PRICE

www.ingramcontent.com/pod-product-compliance
Lightning Source LLC
Chambersburg PA
CBHW081450170526
45166CB00008B/2383